Copying Machines

Copying Machines
t a k i n g
n o t e s
f o r t h e
a u t o m a t o n
Catherine Liu

 university of
minnesota press
minneapolis
london

Chapter 3 originally appeared as "From *Faux Pas* to *Faut Pas*, or On the Way to *The Princess of Clèves*," *Tulsa Studies in Women's Literature* 17, no. 1 (spring 1998): 123–44; copyright 1998, The University of Tulsa. Reprinted by permission of the publisher.

Chapter 5 will be published in *Dramas of Culture: Between Philosophy and Literature*, ed. John Burt Foster (Evanston, Ill.: Northwestern University Press, forthcoming), and appears here courtesy of Northwestern University Press.

Copyright 2000 by the Regents of the University of Minnesota

All rights reserved. No part of this publication may be reproduced, stored in a retrieval system, or transmitted, in any form or by any means, electronic, mechanical, photocopying, recording, or otherwise, without the prior written permission of the publisher.

Published by the University of Minnesota Press
111 Third Avenue South, Suite 290
Minneapolis, MN 55401-2520
http://www.upress.umn.edu

Library of Congress Cataloging-in-Publication Data

Liu, Catherine.
  Copying machines : taking notes for the automaton / Catherine Liu.
    p.   cm.
  Includes bibliographical references and index.
  ISBN 0-8166-3502-1 (hc : alk. paper) — ISBN 0-8166-3503-X
(pb : alk. paper)
    1. Criticism—Europe—History—20th century. 2. Deconstruction.
    3. Literature—Philosophy. 4. French literature—18th century—History and
criticism. 5. Robots in literature.   I. Title.
PN99.E9L58 2000
801'.95—dc   2100-008066

Printed in the United States of America on acid-free paper

The University of Minnesota is an equal-opportunity educator and employer.

11 10 09 08 07 06 05 04 03 02 01 00   10 9 8 7 6 5 4 3 2 1

Contents

Acknowledgments   vii

Introduction   ix

1 Doing It Like a Machine   1

2 "What's the Difference?"   21

3 The Princess of Clèves Makes a Faux Pas   49

4 Getting Ahead with Machines? The Cases of Jacques Vaucanson and Thérèse des Hayes   76

5 Don Juan Breaks All His Promises but Manages to Keep One Appointment (with History)   106

6 De Man on Rousseau: The Reading Machine   127

7 Friends: Dangerous Liaisons   155

Notes   183

Index   219

Acknowledgments

I am grateful to the following organizations for their generosity and support. At the University of Minnesota, I would like to thank the Graduate School, the Department of French and Italian, the Department of Cultural Studies and Comparative Literature, the McKnight Land-Grant Professorship, and McKnight Travel Grants.

Grants from the Société des Professeurs Français et Francophones and the French department of the City University of New York also funded various phases of early research and development.

I would like to thank Crystal Eitle for her attention to the manuscript and William Murphy for his support and assiduity.

The other individuals to whom I am indebted are too numerous to name. I would like to thank all those who had a hand in making this book possible. I dedicate this book to them.

Introduction

> I am my own creation.
> —Madame de Merteuil, *Dangerous Liaisons*

> I am one thing, my writings are another matter.
> —Friedrich Nietzsche, *Ecce Homo*

THIS WORK IS A SUSTAINED EXAMINATION of the automaton as early modern machine, and curious ancestor of the twentieth-century robot, who slaves away at the assembly line of being, sustaining the most precious fantasies of our humanity, while entertaining us with nightmares of the treachery of others. In Walter Benjamin's "Theses on the Philosophy of History," the Turkish attire of the automaton is slightly faded and dusty, giving it an air of obsolescence, quaintness, and disrepute.[1] On the one hand, the chess-playing automaton is considered an allegory for a relationship with the "magical" operations of ideology; on the other hand, it offers us an invitation to a historical perspective, and an elaboration of the ideological and identificatory impulse in our understanding of history and literature. "It was a false automaton which by itself caused more talk than all the others put together, and also acquired a European reputation. This was Baron von Kempelen's Chess-Player."[2] The von Kempelen automaton chess-player was indeed attired in a vaguely Oriental way, and after being presented at the Viennese court around 1783, circulated through Europe in less-than-illustrious circles. From the very beginning, there was a suspicion that it was what historians of the automaton Alfred Chapuis and Edmund Droz call "false," but it provided Walter Benjamin with a complex figure of the struggle between historicism and historical materialism. The historical

Introduction

materialist resists identification, but in so doing she must struggle on two fronts, in the strategic game of chess with an external, visible opponent, and with the "contraption" or game of mirrors that must be operated in order to conceal the complexity of her theoretical investments. Between impatient anticipation of redemption and belatedness in relationship to loss, the automaton of historical materialism is crystallized as an image that is not determined by being but, rather, emerges as a flash of insight that disrupts the very temporality of the event.

The "now" of academia in which I came to intellectual maturity is distinguished by a generalized sense of having overcome the recent past. This is the intellectual event to which I can bear witness: there has been a sense that the theoretical investments of the past few decades had been dissolved. This sense is based on a progressive, additive notion of history in which the present always seems so much smarter than the past. A Benjaminian critique of such an attitude of complacency is founded on an analysis of the historical event in terms of rupture, revolution, and redemption. Criticism itself is only possible if it can think through and leave room for radical discontinuities between past and present, history and materialism.[3] Benjamin reminds us never to take for granted the leap of cunning that allows the theorist of historical materialism to conspire with images of the past in order to redeem the present. Benjaminian recognition must be differentiated from historical identification. The identificatory impulse in any reading of history or literature is more than simply narcissistic; it is first and foremost political. In a purely specular relationship with historical material, we always see ourselves in the victors.[4]

In the work that follows I will focus on the specificity of the metaphor of automaton and the machines as mechanical double of the human being in ancien régime France, because it was there and then that the singularity of the curious and useless machines makes history. The automaton is a preindustrial, nonproductive machine that still has a relationship to the machines of the Ancients. It inspires both automation and mass production, but it ends up as one of the Industrial Revolution's mechanical victims. Its obsolescence is guaranteed by the virtual impossibility of its mass reproduction. The suspicion of the machine, as it is manifested in critical, literary, and philosophical works that we will examine, is not something that can be easily undone, nor should it be. The Enlightenment promulgated two machines: one represented an image of soullessness; the other was sublimated into the very structure of its own ambitions of encyclopedic and instrumentalizing systematicity. Julien Offray de La Mettrie mobilizes an ironic concept of the machine as a limited model for the human being, but he rejected the mechanistic reason as a model for thought when he offers in its place the trope of irony and doubled meanings as the most advanced form of thinking.[5]

Benjamin's automaton of historical materialism is a predecessor to what I call in the first chapter the "theory robot," a figure criticized by journalists and humanists alike as nothing more than an ambitious nihilist. The marionette/automaton/machine is an image solicited by many of the authors to describe ideology itself, but in Paul de Man's work the machine functions as only one part of the system of allegory and irony that is a condition of every attempt at figuration and representation. De Man's materialism is something that I would like to take seriously, in both positive and negative ways. Although his work allows for a sustained critique of simple identificatory impulses in reading, the limitations that it places on itself with regard to psychoanalysis create a detachment that invites transferential aberrations on the part of his readers and students in terms of his asceticism, his teaching, and literary theory in general. Only a psychoanalytic intervention can take into account the way in which his work and person have produced such great loves and hatreds in the recent history of literary theory. This psychoanalytic method abjures a simplifying analogical relationship between models of subjectivity and models of either readership or spectatorship, and is concerned primarily with the disruptive effects of projective mechanisms.

If detachment can be fascinating, the stupidity of the automaton is also hypnotizing, literally mesmerizing in its idiotic repetition of anthropomorphizing movement. What this hypnosis produces is a nonthinking repudiation or total acceptance. The automaton allegorizes the problem of thinking confronted by nonthinking: this encounter takes place, however, in a flash of misrecognition. Thinking thinks that it sees its Other in the passive, glassy gaze of the automaton, but it only sees itself, unable to think outside of specular models. The prevalence and virulence of ideological and self-reproducing systems of judgment produce an indifference that functions as both resistance and repetition. This ambivalent formulation has in the name of progress suppressed critical thinking inside and outside our institutions of higher learning. The figure of the automaton mediates the representation of a catachrestic imperative: how has the Enlightenment represented the machine as its infernal Other, while at the same time adopting a principle of mechanical reason to justify the giddy optimism of its expansionist project? Only historical materialism can answer this question.

Introduction

One of my goals in engaging upon this project has been to expand the range of references available to scholars and students of French literary studies. On a different scale, however, I have also hoped to address the various symptoms of resistances not only to theory but to thinking as well. In theorizing the genealogy of the machine (and by proxy, the automaton), I have hoped to redeem them from unreflected prejudice. Following on the footsteps of such a redemption is a sustained critique of a mechanical humanism that has limited the scope and breadth of academic work.

In the first chapter, I examine the mechanical prejudice as a function of seventeenth-century ethics in a reading of La Bruyère's description of the courtier, and I do so in the context of Paul Bénichou, sociologically oriented critic of the French Classical Age. His resistance to a constellation of theoretical interventions that took place on the French intellectual scene at the end of his illustrious career is considered in the context of a discussion of La Bruyère's own critique of worldly ambition. In the second chapter, we continue to look at the rhetoric of extreme forms of resistance to literary theory. In such a context, it is necessary to address the theorization of "difference" as it has been raised by Paul de Man and Jacques Derrida. The question of difference is ridiculed by David Lehman in *Signs of the Times,* the supposed exposé of deconstruction and Paul de Man. Lehman's wholesale dismissal of the question of "difference" as a challenge to hermeneutics is brought to bear upon a reading of Ridley Scott's *Blade Runner.* In the film mechanical replication and the suppression of difference are addressed in an atmosphere where the relationship with the Other is mediated by the suspicions and violence of film noir. We return to Descartes in order to understand how a rejection of metaphysics takes place in Scott's treatment of the instability of differences between human beings and their technological doubles.

In chapter 3, we take up the question of sexual difference, raised in the context of *Blade Runner,* but this time in a reading of Lafayette's *Princess of Clèves* and of the contemporary criticism of this novel. This novel has been an object of contention and as such has produced a fascinating body of criticism: the question of Nietzsche's will-to-power is brought to bear upon a feminist reappraisal of the princess's enigmatic decision. The self-discipline and the taming of the passions that the novel represents set the stage for a confrontation between love and reason. I try to open up feminist criticism to a more complicated account of ambition, will to power, and self-representation. My analysis focuses on moments of disjunction as literary occlusions of intentionality. I have tried to revisit the debates

Introduction

around this novel in order to revive a discussion of the literary object and its nonphenomenological complexity.

In chapter 4, we examine closely the career of Jacques Vaucanson, an automaton maker of the eighteenth century. In the context of his worldly successes, we look at the progress of technological innovation, scientific progress, and the notion of the contract. His success story is contrasted with the fate of Thérèse de la Poupelinière, a young woman of obscure origins who rose to prominence like Vaucanson in the rapidly changing Parisian scene of the 1740s. Her fall from grace and his acceptance by the Academy of Science are considered in light of the questions of arrival, social mobility, and the fate of women in these years of Enlightenment foment. In chapter 5, we return to the seventeenth century in order to gain insight into the nefarious, mechanical performativity of Don Juan's mastery of seduction. His technique is read as a linguistic one, and one that is fully theorized by the relationship between debt and accountability. That a deus ex machina must be mobilized in order to destroy the libertine destroyer of superstition creates the ground for theatrical ironies within the confines of a strictly policed medium. Here, I also take on an examination of twentieth-century, existentially inflected criticism of *Don Juan,* and the theoretical engagement with this play, from Søren Kierkegaard to Sarah Kofman.

Chapter 6 is a reading of De Man, Rousseau, excuses, and a continuation of the discussion of performativity in the context of the machine. De Man's reading is extremely valuable, also strikingly limited with regard to the question of sexual difference in the context of a rereading of Rousseau's confessional compulsions. Chapter 7 presents a reading of *Dangerous Liaisons.* Here the novel itself produces both pleasures and excuses for those pleasures, even as it describes the machinations of Valmont and Merteuil to dominate their peers, and to master each other. As a rewriting of other novels of worldliness and initiation, it offers stunning examples of dissimulation and self-deception. As the threat of Don Juan's libertine and secular reason grows, the Enlightenment produces a symptomatic cult of sentimentality that attempts to seal the breach opened up between pleasure and love. The detachability of eroticism and love is discovered by Merteuil, who, I argue, is the Absolute Philosopher. The fact that she is a woman raises the stakes in her search for a method to master her feminine passions. That philosophy must be restrained in its pursuit of truth, especially when it comes to the question of women and sexuality, is made legible in this literary figure. Valmont and Merteuil participate in disseminating mechanical prejudice: they often dismiss their victims as stupid by means of a mechanical analogy. This makes them distinctly modern and familiar to us, especially after we have examined so many examples of such derogatory characterizations of the machine in contemporary criticism.

Many of the concerns addressed here can be identified with issues that became pressing in the penultimate decade of the twentieth century. It seems more and more obvious that no matter how much we would like the debates of the recent past to disappear, they continue to haunt us by violating the periodizing boundaries that have been set up in order to keep us all in our places. For some of us who came of age in the eighties, inspired by a certain kind of theory now marginalized as "high," we have survived only by refusing to see ourselves either in the lost glory of the past or in the pious complacency of the present. This work is testimony to such survival, and as such it is an ambivalent object. I hope that this book participates in the urgent thinking through of the eclipse of both literary theory and the literary object.

Chapter 1
Doing It Like a Machine

IN *ALLEGORIES OF READING,* PAUL DE MAN describes the grammar of a text as functioning "like" a machine.[1] What are some of the consequences for literary interpretation of this comparison? This is a question that this chapter hopes to answer in a provisional manner; the arguments presented here will be worked out in the rest of the book in a series of readings, of both primary texts and secondary debates. De Man's point is rather modest at first glance: he suggests that the grammar of a text labors tirelessly to produce interpretative possibilities that are relatively impervious to the intervention of the author's intention or the interpreter's good will and intelligence. In confronting the semantico-grammatical autonomy of a text, de Man seems to inevitably proceed toward a showdown with something idiotic about writing and reading. A particular and even peculiar textual idiocy can take the form of "referential detachment, gratuitous improvisation ... [and] the implacable repetition of a preordained pattern." As others have made clear, de Man's power as both critic and pedagogue was based in large part on his own fascinating ability to reproduce a high degree of detachment in intellectual, pedagogical, and political relationships.[2] De Man's use of the figure of the machine in describing the grammar of a literary text, and his characterization of language as functioning under the conditions of referential detachment, precipitated a crisis in literary studies that we are only now beginning to address.

Throughout this book, de Man will be read as a symptomatic figure of the institutionalized study of literature. In order to be able to better account for the impact of his work on the crisis in literary studies, we are going to read him with and against a tradition of literary criticism that has been mostly resistant or impervious to his work. The trajectory of this analysis will lead us back to a theoretical genealogy of the machine, and especially to the figure "like a machine." It is the machine, after all, that recurs in de Man's own work as a crucial figure and then reappears in the criticism of his work, and the work of deconstructive critics in general, as an embodiment of the infernal principle of both repetition and detachment. For critics of de Man, if the text cannot be compared to a machine, de Man and his disciples should be.

De Man's detachment implies an act of reading distinctly at odds with the ideas of many literary critics. The conflicts within literary studies that de Manian intervention produced are based on the fact that critics can no longer proceed as if there were a consensus about the object of their study. The rift appears immediately when critics on one side of the divide use the very figure of the mechanical or the machine as that which is antiliterary, and not just nonhuman. The Cartesian prejudice against the machine is difficult to overcome, but most recently this particular myth of the Enlightenment (based on the idea that the human being and the machine were to be differentiated in a radical way) has been redeployed by critics of literary theory who defend literature in the name of its singularly nonmechanical qualities. That de Man could so offhandedly and so lightly dismiss the grammar of a text as being "machine-like" is, for theory's critics, only the beginning of his error.

The consequences of not addressing the implicit denunciation of the machine have far-reaching ideological consequences that have yet to be examined. Let us take the case of leftist critics of the midcentury like Paul Bénichou, whose work on the French seventeenth century must certainly be regarded as a strong political intervention in the tradition of French literary studies of the classical era. His work in Old Regime French literature was groundbreaking insofar as he was able to offer a powerful alternative to the airless nineteenth-century historical accounts of the period. His *Man and Ethics: Studies in French Classicism* offers a powerful reading of classical authors like Molière, Corneille, and Racine in light of the radical social changes taking place in seventeenth-century France.[3] The struggle between feudalism and modernity, Bénichou shows, provided one of the determining conditions of religious thinking and literary representation. If his work is opposed to the monument of French literary criticism that is Guy Lanson's *Histoire de la littérature française* (1894), it is because Bénichou revises Lanson's notion of history by introducing the notion of social class and class struggle into Lanson's grand historical narrative. Bénichou is in many ways the first historical materialist to take up a careful analysis of this century of the consolidation of not only the French nation-state, but French culture in general.

It is very surprising that in *Le Statut de la littérature: Mélanges offerts à Paul Bénichou*, a Festschrift published in his honor in 1982, there appear a number of instances when both Bénichou and his followers feel the need to defend themselves and their work against the work of the "theorists."[4] What is it about the literary theorists that a critic like Bénichou does not like? The answer to this question is not a simple one, and it demands a careful examination of his assessment of the activity of the literary critic in his "Réflexions sur la critique littéraire." I shall cite him at length:

> We are not manipulating bodies or machines; we listen, we interpret, we confront signals and wills. Our concern is to do it with the least amount of error possible; our criteria for truth, inevitably approximate, and rarely providing for certainty, demand to be handled with care and rigor, and they must at the same time be protected from all fantasies that claim to have forgotten man.[5]

At first, in Bénichou's careful treatment of the notion of certainty, one could identify a kind of compatibility with de Man's literary theory. Upon further reflection, however, any kind of dialogue between the work of the two critics disappears. While literature, according to Bénichou, must be handled with care, literary critics are not allowed to use their hands—"we are not manipulating bodies or machines"—or any parts thereof. Literary critics have to be all ears. A literary text is not a machine, and literary critics are not mechanics. Bénichou's very polemical "Réflexions sur la critique littéraire" is haunted by the ghost of a Cartesianism that insinuates itself as the force of reason. Bénichou describes his "historical materialism" as a kind of sociological criticism that abjures all method, but his distinctly antimaterialist approach to reading the literary text is obscured by the almost unrecognizable denunciation of the machine as a figure of an antiliterary matter. In his historical materialism, the literary text itself is what is dematerialized as a series of signals and signs.

The way in which Bénichou describes his approach to the literary text, however, resembles Lanson's insofar as both critics conceive of the literary text as an expression of the author's "individuality." Bénichou defends this idea against what he and his followers call theory.[6] An example of this line of defense can be found in Jean Molino's "Sur la méthode de Paul Bénichou."[7] In this essay, Molino takes aim at what he contemptuously calls "le caravanserai" of the "Supreme Theorists" ("Suprêmes Théoriciens"), and he describes Bénichou as a skeptic whose prudence makes him erase all reference to any system, or system making. It is in this act of self-erasure that we can see the traces of a ghost-writing, dictated by Descartes himself. Bénichou's arguments are shaped by the rhetorical figures of a systemic Cartesianism, the first sign of which is obvious in his explicit refusal of methods and methodologies as so many forms of prejudicial approaches to the literary object.

Bénichou describes Marxist, psychoanalytic, and structuralist critics as having strayed from the path to the truth by taking bewildering and disorienting detours ("pérégrinations dépaysantes"). Bénichou, on the other hand, is always trying to orient himself toward the "light" of truth. The best way to approach the truth of literature is by following an implicitly Cartesian itinerary, that is, by refusing all method and abandoning all previously learned systems as prejudice in order to confront the literary object in all of its luminosity: "It was necessary to forget the system as such and work without prejudice while orienting oneself in the chosen direction."[8]

This Cartesianism claims the truth as proper to a hyperreasonable, invisible nonmethod. The critics who have gone astray are hallucinating (they are the ones susceptible to chimeras). Perhaps they are in fact dreaming, while they are manipulating a phantom text, all the while being duped by an Evil Genius under whose spell, like drugged mechanics, they tirelessly travail. They are heavy-handed in their approach to the text: they are the ones who manipulate and mishandle. Perhaps they even take a too-intensely hands-on relationship to reading and thereby forget to listen and watch for the signals and the voice of the author's ghostly, but tenacious, will. Perhaps handling is manhandling here: the critic's job is one that can only be performed with no hands.

The text is a fragile vehicle that transports the "signals" of the author's will, and not some stupid, empty machine. Bénichou's Cartesianism is the promulgation of a method that aspires to set itself up as the highest order of reason: it is the only thing that is immune to ideologies that seek to deform and manipulate the literary object. Other methodologies violate the sanctity of the literary object by producing an ersatz text whose reading serves only the purposes of ideology.[9] Bénichou's arguments appear extremely reasonable: they mime the very discourse of reason itself. Bénichou, like so many French academics trained in the art of the résumé, so successfully incorporates the rhetoric of Cartesian reason in his literary criticism that he appears to be defending reasonableness itself against the deformers and reformers of literature, who only seek to further their own causes at the expense of the text itself.

Predictably, in the Cartesian system the machine appears as the figuration of all that literature is not. In the metaphorical language of Bénichou's own text his antimethod reveals itself as a genealogy of morals that deploys itself against the machine. Literature is organic, evolving, saturated with the will to communicate. The machine is a mechanism of compulsive repetitions; it is hermetic and autistic; it is dead matter. As such, it is an ugly harbinger of death, and on top of it all, it is vulgar and stupid. After having shown us how others have gone wrong in the study of literature, Bénichou has shown that it is absolutely necessary to apply the "handle-with-care" label to literary works. To follow his arguments to their logical conclusion, however, in order to avoid any kind of "manipulation" of literature, one should simply keep one's hands to oneself. With the intervention of the hand, the danger of manipulation and abuse arises.

Textual harassment or interpretative molestation is only one of the crimes of which a bad critic might be accused. One of the most graphic ways in which this harassment/abuse can take place is in the critic's careless mishandling of the fragile container of an author's will. The critic must be ever vigilant about securing the truth of the author's presence in his work, which is a work of the spirit and not of the body. Instead of manipulating corpses and machines (for critics are neither undertakers nor mechanics), we must watch for signals and "wills" (critics are closer to the model of the psychic or the clairvoyant). The good critic does not force or manipulate the text; he peers into it as one would into a crystal ball. The devaluation of the body/machine in favor of the mind/*esprit* sets the scene for a situation in which the work of the critic begins to resemble the work of a nineteenth-century spiritist. One waits for signals from the dear and departed, on the watch not for signs (for that already is too material) but for signals of the absent one—a tapping on a table, the tipping of a painting. This description of what might be called reading is the result of Bénichou's applied historical and sociological materialism.

The handling is mishandling, abuse, a manipulation, and violation. The intervention of the hand in Bénichou's account takes place as an act of clumsy prestidigitation. Manipulation is always what is at risk, when we take a hands-on approach to reading. This is only one aspect of the theoretical intervention in literary criticism that Bénichou finds reprehensible. Paul de Man has been accused of many crimes, only one of them being the abuse of literature. There are those whose objections to his work take the form of a purely biographical condemnation: more sophisticated critics like John Guillory find that his techniques of literary analysis exist in an unconsciously mimetic relationship to bureaucracy.[10] According to Guillory, de Manian detachment permits for a kind of institutional complicity, but it could be argued that de Manian detachment demystifies a certain critical attachment to metaphysical categories, especially when it is deeply encrypted in the writings of the most interesting historical materialists. De Manian procedures of reading thus amplify, exacerbate, and also clarify at the same time a very precise conflict that takes place within the field of literary studies and in institutions of higher learning.

"The machine is like the grammar of the text when it is isolated from its rhetoric, the merely formal element without which no text can be generated. There can be no use of language which is not, within a certain perspective thus radically formal, i.e., mechanical."[11] For de Man, the radically formal aspect of all texts is compared with the machine and its mechanical operations. All texts follow a grammatical logic that produces certain inter- and intratextual combinations having nothing to do with the will of their authors. Reading for such moments of radical formalism is perhaps reading like a mechanic, but it also gives an entirely different place and weight to the question of uncertainty. In fact, the figure of the machine in the text liberates writing from a relationship of instrumentality vis-à-vis the spirit—the *esprit*. Bénichou moralizes against such reading by denigrating both the hand and the machine—catachrestic figures that we will return to again and again in our readings in the Old Regime in order to understand how notions of "working with" literature and "working in" literature function.

Bénichou's resistance to theory notwithstanding, it is he who gives us access and real insight into a reading of French moralist La Bruyère. In the work of the latter, however, we will encounter a denunciation of the machine that draws a relationship among the categories of the automaton, the courtier, and the idiot. These three terms function as predicate nouns in certain sentences in La Bruyère's *Les Caractères,* and their mode of descriptive condemnation leads us back to what we could call the mechanical prejudice. According to the historical perspective, by the second half of the seventeenth century, the waning of feudal power is the material condition of moral pessimism. As Bénichou emphasizes in his discussion of Pascal's thinking, the "automaton of habit or custom" has very debilitating effects of habit on the autonomy of human reason. Materialism, as invoked by the automaton of habit, deforms the integrity of human decision. The degradation of the moral life of human beings has to do with the communicability of matter (automaton) and mind: "The automaton can only influence the mind because there is constant 'communication' between the bodily machine and thought."[12] It is in this way that human decision and moral life are compromised, and the figure of the force of this compromise is the automaton. In Pascal's and La Bruyère's seventeenth-century moralizing, the automaton and the machine become conditions of critical thinking itself.

In La Bruyère's taxonomic system, described by Louis Van Delft as a kind of classical anthropology, the situation of each character in his or her appropriate place produces a moralizing topography.[13] The place of the automaton is connected to the places of both the courtier and the fool. Mechanical repetition is the very image of moral and spiritual failure. The machine is a crucial figure in La Bruyère's critique of stupidity, complicity, moral failure, blind ambition, and self-interest: the futility and idiocy of the machine, the courtier, and the fool form the ground on which are founded the positive values of agency, intelligence, and probity. The machine works in this text by posing in various places as the very figure of tautology: it is censured because it produces nothing but more self-serving mechanical motion.

The explicit intention of La Bruyère is pedagogical, and this intention is framed in an imperative that, according to the author, should apply to all linguistic interventions: "One should not speak or write except to instruct; and if it should happen that one pleases, it is not necessary to regret such a thing, if pleasure serves the acceptance of the truths that are meant to instruct."[14] Jean Starobinski's analysis of the conventional formulae of sociability emphasizes that a certain concerted suppression of aggression is necessary for the pleasures of civilized social life.[15] It is, however, what he calls the radical aestheticization of the social field that produces a certain anxiety about the authenticity of any gesture.[16] La Bruyère attempts to put pleasure in its place by appealing to a higher order, that is, to the exigencies of truth.

Pleasure may be produced as a by-product, but producing pleasure should not be the intention of the writer. However, this pleasure, once produced, can also be useful in mitigating the unpleasant effects of a truth that is difficult to swallow. Pleasure as a secondary effect is acceptable; the author is not obliged to deny its place in his work. Pleasure allows truths to insinuate themselves more effectively because it allows thoughts easier entry into the mind of the reader or listener—in a sense, pleasure performs a courtier-like function, the only purpose of which is the lubrication of certain channels of the transmission. However, La Bruyère thoroughly condemns orators and writers who write only in order to please: "The orator and the writer cannot overcome the joy of being applauded; but they should blush at themselves, if by their discourse or their writing they had sought praise."[17] It is difficult not to be pleased that one pleases, but if one looks only to receive praise and give pleasure, then one's work is without any value—it is shameful. Pleasure cannot be the final or primary condition of writing or speaking. It is necessary to write and speak of unpleasant truths at the risk of displeasing. If in the process one manages to produce pleasure, it must happen despite oneself. La Bruyère leaves open the possibility that pleasing one's readers can happen, sort of as a side effect of one's explicit will to instruct. But the paradox of pleasure is that it is only acceptable when it is produced as an unintended secondary effect. When it is the result of explicit intention, then it should only make us blush with shame at our own pleasure at pleasing. Pleasure of any sort in relation to discourse and writing is invalidated if there has been a will-to-please. Once producing pleasure for pleasure's sake is possible, then an infernal mechanical process is set into motion whereby the pleasure that the author experiences in pleasing his audience motivates him to only please: the will-to-please would then usurp the place of the will-to-instruct, and everyone involved would be blushing with shame (and pleasure).

Jules Brody suggests that *Les Caractères* distinguishes itself by proposing an examination of a new reality—an order of superficiality that the text explores by metabolizing and imitating its very form: "From this time on, because the world is nothing more than surface, 'the real substance' of writing will be 'this superficiality' by means of which style makes itself homogeneous with its object."[18] The object of representation here is superficiality itself; in the writing of superficiality, it is the style that will mime the superficiality of the world that the writing seeks to represent. Writing about superficiality means writing superficially: the metaphor of surfaces without depth, however, can also be understood as leading toward the "literal" surface of writing itself. Writing would not be possible if it were not possible to imagine that signs could be imprinted on surfaces without depth—and that meanings could proliferate horizontally rather than vertically. Van Delft proposes that we think of this as a typographical model.[19] Each character is like a letter of the alphabet: the writer produces his own typeface that becomes legible to a reader after a period of familiarization. If we think of these two analyses of La Bruyère together, we can understand the text as both topographical and typographical. Surfaces, after all, are what offer themselves up to be read, and when we are accused in English of reading too much into things, we are accused of having punctured the surface of a text with our overreading. If texts, people, and meanings can be described as having "depth," we must think of that depth as being one that will only be acknowledged, not plumbed. It is still, in so many circumstances, considered better to be deep than shallow.

The topography of typography offers up a model of reading that we will try to elaborate. Here is a copying machine—writing that traces in its style the object of representation—that makes sense of what it is copying by creating a new typography. One of the subjects of superficiality that La Bruyère hopes to isolate from the topography of the court is, of course, the courtier. It is not insignificant that the courtier and the idiot ("le courtisan" and "le sot") are the two characters who are described in mechanical figures.

The courtier of the seventeenth century is no longer Baldassare Castiglione's man of learning whose presence at court is justified by the honorable desire to put his learning and wisdom at the disposal of his prince in order to serve the city or nation-state. He is the craven man of ambition who has no inherent qualities or innate abilities. He is only lubricious: he is all will-to-please, and his pleasing does not only produce for him the pleasures of being praised, but also the pleasure of seeing his own interests advanced at court. He serves no one but his own cause, yet his futile machinations lead him absolutely nowhere:

> With a watch, the gears, springs, and movements are hidden: nothing appears but the needle, which advances imperceptibly and completes its turn: this is the image of the courtier, and it is all the more perfect insofar as after having traveled his path, he often returns to exactly the same point from which he began.[20]

The courtier's ambition, his driving force, makes him absolutely predictable: his character is reduced to that of a clockwork mechanism whose movements are completely determined in advance. While the courtier believes himself to be advancing, his movement only marks the passage of time. The courtier must always be in movement, but every move is absolutely calculated and calculating. Although he may believe that he is making a kind of linear progress toward a higher position at court, amassing more power and influence and prestige, in the image presented by La Bruyère he is, in fact, turning around in circles, in circles that mark the passage of time until death and Final Judgment, when all his efforts will be shown to have been in vain. It is in the double meaning of the verb *s'avancer* that the irony of the image of the courtier rests. *S'avancer* describes both the passing of time and the improvement of one's position. The courtier believes that he is getting ahead ("qu'il s'avance") and he is, but only because he is marking time on the face of a watch:

> The image of the moving hand is, as Bergson has shown, essential to the representation of the non-qualitative time of the mathematical sciences. This is the context within which not only the organic life of man is enacted, but also the deeds of the courtier and the action of the sovereign who, in conformity to the occasionalist image of God, is constantly intervening directly in the workings of the state so as to arrange the data of historical process in a regular and harmonious sequence which is, so to speak, spatially measurable.[21]

Walter Benjamin remarks on the juxtaposition of mechanisms of keeping time and machinations of the courtier in *The Origin of German Tragic Drama*. In his description of the temporality of court life, however, the potential intervention of the sovereign is invoked. The despair and melancholy of La Bruyère's mechanical courtier are founded on the meaninglessness of time for the ultimate Clockmaker of the Clockwork Universe, God himself. For God, time is infinitely expandable (He is the infinite) and infinitely contractable (passing time has no significance for Him). Marking time for God is ultimately a futile activity.

In Benjamin's take on the Baroque courtier, it is the potential action of the King, the stand-in for God himself, that places an absolute limit on the courtier's actions. The sovereign can always reset the clock, as it were, slow down and speed up time and force history into a logic that will serve him. The courtier's efforts to curry favor with the sovereign and advance his own cause will amount to nothing with one simple gesture of refusal and rejection: for to occupy his place, many others are waiting. The courtier plays this serious and melancholy game ("la vie à la cour est un jeu sérieux, mélancolique"), and it is his actions, his movements that keep time at court, that guarantee the numbing regularity and monotonous rhythms of this entirely ceremonial life: he is waiting for the intervention of the sovereign, who will either consecrate his efforts or send him into oblivion.[22] According to Benjamin's analysis, the sovereign is the Cartesian God of the political world: "In the course of political events intrigue beats out that rhythm of the second hand which controls and regulates these events" (97).

Doggedly faithful to obeying the rules of a game he cannot hope to win, the courtier is a creature of discipline and despair—a machine, then, whose movements copy the marking of time itself. It is in both his own disciplined regularity and his understanding of the predictability of "human nature" that the courtier is the object and subject of intrigue at court. In German tragic drama or mourning-play *(Trauerspiel)*, the courtier is the character who plots and sets plots into motion, for he is the "sovereign intriguer" who "is all intellect and will-power" (95). In the inevitable futility, however, of his machinations, we see represented the impotence of intellect exercised in the court of any sovereign. For Benjamin, the courtier is more a figure of tragedy than a figure of evil: he is two-faced. It was in Baroque Spanish theater that this dual nature was fully realized: "[The courtier is] the intriguer, as the evil genius of their despots, and the faithful servant, as the companion in suffering to innocence enthroned" (98).

It is the combination of intellect, insight, and absolute dependency that makes the courtier such a fascinating character, one of the more neglected and important figures in Benjaminian thought. The courtier is mediator and intermediary: even in simply marking time, he is acting as a medium by which the movement of celestial bodies is made legible to man. After all, it is important to learn how to read clocks and tell time: it is not an innate ability. The courtier is both messenger and transmitter. He embodies communication from a distance: his activities produce the infinite distance that must separate the absolute monarch from his subjects. As in the portrait of the courtier of La Bruyère, the courtier in Benjamin is the one whose failures are almost inevitable. Too dependent on chance and the whim of the sovereign, the fate of the courtier is never guaranteed by the extent of his renunciation. In the portraits of vile courtiers in La Bruyère, the courtier is willing to trade dignity, virtue, honesty, integrity, and happiness for a chance to succeed at court. For the seventeenth-century moralist, this is absolutely reprehensible behavior; in Benjamin's allegorical analysis, the courtier can be understood as a figure whose sacrifice embodies a kind of secular saintliness and a kind of heroic renunciation. The courtier as communicator calls up the image of compromise itself: for it is, after all, the communication of base materialism (automaton) with the mind *(esprit)* that forms the very condition of Pascalian compromise.

One of the great courtiers in literature must also be mentioned here as a figure who crystallizes the drama of this life of intellect's dependency on despots, the count Mosca of Stendhal's *The Charterhouse of Parma*. In this novel, Mosca's intelligence is his tragedy. Much feared, he is trapped into defending a vain and stupid prince from the violence of equally vain and stupid revolutionaries. The failure of the courtier allegorizes something about the failure of intelligence before the intervention of despots as well as the revolutionaries who would depose those despots. Mosca's fate demonstrates that the triumph of intelligence implies something about the impotence of the most experienced intriguer, the most experienced politician, the most sensitive lover in the face of the stupidity of a despot (the Prince of Parma), the idiotic zeal of the revolutionary (Fabio Conti), and the ingenuousness of the handsome young man (Fabrice himself).

The courtier is the figure of intrigue and conspiracy: for La Bruyère, however, his mechanical regularity betrays his radical duplicity. To be a "honnête homme" or "un homme d'esprit," one must be capable of being spontaneous, unpredictable, "inégal" in the sense of irregular. This spontaneity is founded on the possibility of linguistic invention, a quality lacking in machines and animals. In *Discourse on Method,* Descartes proves the absence of the capacity to reason in animals and machines by way of their inability to produce a spontaneous linguistic formulation. Descartes writes in his correspondence with Morus that the language used by human beings is "the unique sign and only mark of thought concealed by the body."[23] The signifying power of the word is related to its power to represent thought contained by the body. Even the most intellectually limited of human beings is able to arrange different words in such a way as to communicate their thoughts. That the word represents presence of thought, presence of mind, and presence of soul anchors the differences between human being and animal-machine. Jules Brody and Michael Moriarty both demonstrate with great effectiveness that in La Bruyère, *parole* is susceptible to all sorts of disruptions, and that these disruptions are significant in the context of the Cartesian line drawn between animal-machine on one side and human being on the other. When referentiality is destroyed and speech is no longer attached to thought, speech becomes physical, bestial, and mechanical: "In order that speech can serve to distinguish man from the reign of the animal, it is necessary that it be the function of the intelligence and the will of the one who speaks and that it be recognized as such by the one who listens."[24]

The referential power of language, however, is entirely vulnerable to disruption—at least in La Bruyère's application of Cartesian differences in his moralizing project. In the case of the courtier, the referential power of his word has been completely undermined by the power of his ambition:

> One can no longer expect candor, honesty, equity, favors, service, good will, or firmness in a man who, having devoted himself for a certain time to the court, secretly wants to make his fortune. Do you recognize him by his face or by his conversation? *He no longer calls things by their names;* for him swindlers, impostors, idiots, and impertinents no longer exist: the one about whom he says what he thinks is the same person who might prevent him from getting where he wants to go: he thinks ill of everyone but speaks no ill of anyone.... he possesses a sad circumspection in his behavior, and in his conversation there is an innocent but cold and forced playfulness. He unleashes torrents of praise for the things done and said by a highly placed man who is in favor, and for everyone else, he is afflicted by pulmonary dryness.[25]

Because the courtier no longer says what he thinks, he has renounced linguistic referentiality. Nomination no longer works in his signifying system, because his word no longer signifies his thought. By refusing to call things their names, he rejects the designatory power of linguistic material, but in doing so he condemns himself to being able to signify only one dominant inner motivation, which is not so much a thought as it is a blind instinct—ambition. This referential breakdown is described by Benjamin's analysis of the linguistic breakdown in the language of allegory and *Trauerspiel*. Howard Caygill writes in this context, "Language is reduced from the communion of expression and signification to a mediation between object and mind; between man and man.... The word in transition, the linguistic principle of *Trauerspiel*, is an expression of the rupture between original expression and signification."[26] As Benjamin points out, the increasing worldliness of the Counter Reformation existed side by side with powerful religious institutions and aspirations; this contradiction produced a particular set of linguistic and political constraints on writing and thinking: "The only consequence could be that men were denied all real means of direct expression."[27] Like the courtier, the word is reduced to a ceremonial role of mediation and intrigue. The power struggle in which the mediator engages is identified as a void or a nothingness by the moralist.

According to Barthes, La Bruyère's *parole* points to something troubling: the word of the courtier is there, pointing at nothing—nothing, that is, but his ambition. The uttered word of the courtier is a movement calculated in order to further his own interests. Moriarty has called this particular relationship to language "a linguistic pathology."[28] If the purely formal use of language can be defined as a pathology, then this affliction might well become a chronic condition of the court itself: the speech of the courtier and the idiot becomes attributed to a kind of brute physicality, and a total atrophy of referentiality. La Bruyère describes the conversation of the man of the court as an imbalance of the flow of humors: on the one hand, he cannot master the uncontrollable torrents of praise that burst out of him in the presence of the favorites.

In the case of the idiot, however, his absence of intelligence dooms his word to a different yet similar form of emptiness:

> The idiot is the automaton, he is a machine, he is a spring; a weight dominates him, makes him move, turn, always in the same direction, with the same regularity; he never betrays himself; whosoever has seen him once has seen him in every instant and every period of his life; he is at most a lowing steer or singing blackbird: he is fixed and determined by his nature, and if I dare say so by his species. That which appears the least in him is his soul; it is not active, it does not exert itself, it is at rest.[29]

Now the fascination of such a picture of the world is that it spans two methods of taxonomy: it rests on the Cartesian differentiation of animal and human while at the same time drawing endless analogies between the two opposing categories, by assimilating two "species" of men, courtiers and idiots, to machines and animals. The animal-machine is already an analogical monster that allows Descartes, in his letter to Newcastle, to describe the migration of the swallows as the spectacle of so many flying clocks, all perfectly tuned to the changing of the seasons.[30]

If authentic communication is impossible in the case of the courtier and the idiot, something is nevertheless communicated—submission to the dull force of habit, the empty forms of protocol, or the obscurity of matter in its relationship to mind. The courtier communicates his submission to the rhythm of courtly life: his submission in turn communicates his ambition. He is readable as a character and therefore reproducible. He is a type. The idiot communicates perhaps nothing more than his stupidity, but his stupidity allows him to enter into analogical relationships with animals, clocks, steer, and blackbirds. His character is his destiny; once named, he has an important function: he keeps time, along with the courtier, and while the "honnête homme" may be intelligent, you certainly would not want to set your watch by him. The mechanical quality of each type allows for it to be immediately legible in the series of characters that populate La Bruyère's text.

In Furetière's dictionary, among the definitions of the machine is one that is important for our purposes:

MACHINE: used figuratively in the case of things moral, to describe the techniques, the artifice by which one pursues the success of some affair. He set into motion all sorts of springs and machines in order to succeed in this enterprise. This man is vulgar and heavy, he is a machine, he never leaves his chair.[31]

The figurative meaning of *machine* describes a double condemnation: a light-fingered artifice that one deploys in order to succeed in an enterprise in a not-altogether-honest way, and the heaviness and immobility of a weight, the same weight of a dense materiality that is the burden of the not-so-light-on-one's-feet. The spring mechanism and the system of weights, pulleys, and levers that allowed for automatic movement are seen as both deceitful and stupid when compared with the authenticity and the intelligence of spontaneous speech. Returning to Brody: "In considering the machine—animal or automaton—as an imitation, or rather as a bad copy of man, Descartes was able to maintain, in his natural system, the traditional, metaphysical hierarchy which guaranteed human autonomy in relationship to the passivity of the moved thing."[32] More surprisingly, however, in Bénichou's historical materialism, the mechanical prejudice is left intact: a kind of metaphysical communion or communication takes place there. Bénichou depends on theology. In this case, the hidden dwarf in Benjamin's automaton of historical materialism represents a theological respect of the difference between matter and mind.

What allows for the preservation of metaphysically hierarchical relationships in La Bruyère's moralizing system is the attribution of certain machine-like qualities to different "species" of men. A moral hierarchy is thus created, and the autonomy of the *homme d'esprit* is radically differentiated from the submission of the courtier and the idiot to physical laws. The courtier is a bad copy of authenticity and the idiot a bad copy of intelligence. The moralist is one who is unmoved by the venal ambitions of the courtier, and whose linguistic production is free from all base motivations in his aim to instruct in an absolutely disinterested way.

Is it necessary to point to the suspicion under which the machine has operated and how this suspicion has to do with its duplicity and its stupidity? That machine-like, formal aspect of every text—which resides in its adherence to convention—makes it at once legible and deceitful. The censure of the ambition of the courtier in La Bruyère has to be reread in light of his own nominative procedure. If he indeed seeks to name, criticize, and condemn the characters of his milieu in an absolutely evenhanded and disinterested way, there are a few disturbingly formulaic descriptions in his own text that are remarkably banal in their flattery of the powerful. He does not fail to participate in the ritual flattery of the king in the following passage (the allegorical formulation is a familiar one in seventeenth-century France): "The children of the Gods, so to speak, exempt themselves from the rules of nature, they are its exceptions. They expect almost nothing from time and years. In them, merit is always ahead of age. They are educated when they are born, and they attain perfection when the common man is just emerging from childhood."[33]

As Starobinski reminds us, flattery produces a peculiar configuration of the circulation of words and favors: it is the condition of the tyrant's court. That the moralist himself cannot step out of this field is not at all surprising: flattery is no longer a discursive aberration but, rather, the limit of judgment when confronted with an absolute difference in power and wealth between speaker, writer, and interlocutor: "The situation of the poet in a political and social order in which a writer's material resources still depend in large part on the good will of powerful men is therefore quite embarrassing. He is obliged to praise yet at the same time to defend himself against the imputation."[34] Flattery becomes the mark of an ambivalent dependency that threatens to corrupt all authenticity in linguistic communication. Flattery is the courtier's vice.

To describe the precocity of the king's progeny is commonplace, but here the text seems to suggest that the limits of its own pedagogical project lie in the existence of these divine creatures. For if we are to take La Bruyère at his word and heed the injunction that one should neither write nor speak except to instruct, what happens when one is confronted with those born with instruction—those children of the Gods whose very existence defies any sort of pedagogical intervention? All voices should fall silent here, for to address them is impossible according to the first injunction. To address them is to fall into the problem of addressing oneself only to please. Any sort of instructive impulse must be quelled: it is here that the moralist teaches himself the limit of his own intervention. For failing to be able to instruct, he falls into a kind of formulaic flattery that makes him equivalent to the most effusive and hyperbolic courtier. Here the text is shrouded in its own contradictions and, despite the author's best intentions, transgresses the moralist's first injunction. Better, then, to fall silent.

It is in the complicated web woven by the aphoristic style that the problem of intratextual contradiction arises. The honeycomb architecture of this moralizing web of distinct compartments, which are further subdivided into smaller units of highly condensed aphoristic statements, rests on a foundational injunction, which is in turn undone at its very heart by the fall into mechanical flattery. The text is divided against itself: the courtier and the idiot are not so easily named that they stay in their places. The name of man, *l'honnête homme,* is only a name by analogy: the attenuation of the moralizing word in face of the description of the king is disturbing at best. The authority of the moralist betrays itself as a dependent and contingent one at such a moment, but the appeal to absolute authority does not undermine the text's authority—entirely or instantaneously. This flattery is meant for the king but received by a system of intermediaries and mediators. The labyrinth of the court here can be compared to the labyrinth of a giant ear in which the message circulates, hoping to resonate against the innermost tympanum, the ear of the king. Word of mouth is absolutely crucial. As Starobinski writes about Theophrastes' moralizing (translated from the Greek by La Bruyère): "The flatterer is a mouth that first speaks then eats. The theme of the parasite's success in preying on the rich man was incorporated into the classical critique of wealth."[35] The parasite at the table of the wealthy man is transformed in La Bruyère's critique of flattery into the courtier in the tyrant's court. The risks and the benefits in such a relation are markedly higher, and, therefore, flattery of princes becomes more refined and more hyperbolic at the same time. Flattery is based on two complementary appetites: on the one hand, the parasite/courtier is hungry for good meals, favors, and distinction. On the other hand, the powerful and wealthy man, or the prince, is hungry for compliments.

Moriarty points to the image of whispering in ears that occurs throughout La Bruyère's text: the whispering is communication as a spectacle, taking place before an audience in order to address a third party with the message of one's own power or one's intimacy with the powerful.[36] The whispering aspires to a game of telephone in which the message might be carried, ear to ear, until it reaches the final destination, the ear of the monarch himself. Franz Kafka reverses the order of this communication in his short story "Imperial Message," in which the emperor whispers in a messenger's ear, a message for "you," living on the border of the empire, dreaming of the emperor, who on his deathbed sends a message to you with the hardiest of messengers.[37] In the space of these few hundred words, the distance between the dying emperor and the dreaming "you" increases with the unfolding of a labyrinthine space of palaces within palaces, courtyards within courtyards, staircases leading only to more staircases, and all these spaces crowded with courtiers, waiting for the emperor's death. The more the messenger struggles against the crowds, the greater the distance grows between him and you, waiting at your window, at the outer limit of the empire itself.

In La Bruyère, the courtier's torrential flattery is not meant for the ears of a particular interlocutor; the effluvium of praise is produced in order to saturate all ears with his gushings. If his praise is not meant for the ears of his interlocutor, it is meant for the giant ear that is the labyrinth of the court itself. At the center of the labyrinth is a king and the favorite, who in turn is a courtier, who knows how to transmit, censure, and receive messages, and who has the ear of the king. The court is one complicated but internally wired system of telecommunications: it is a space that provides for the transmission of messages even as it undoes the referential power of language. He who resides there actually lives inside a giant ear,[38] a completely wired self-surveillance module.[39] Its intricacies create greater and shorter distances between the subject of language and his addressees. The king, who is the ultimate receiver, is surrounded by mediators and intermediaries: to reach the king is to transmit the message through and by them. Every courtier is a supplicant at the classical Poste Télégraphe Téléphone, which consolidated national boundaries even as it produced postal delivery systems. The linguistic pathology described by Moriarty seems to have to do with the interferences to which all instances of communication from a distance are vulnerable.

If the courtier makes no progress but believes himself to be advancing, crossing the distance between himself and his final destination, it is most important because he is under the illusion that he approaches the center of power. Instead, in the moralist's description, he is just a clockwork mechanism, turning in his circles but never leaving his post. This is his place, the place of the machine, of automatic movement and the autonomic nervous system. He will mark the site of the corruption of referentiality. In judging and classifying both the courtier and the idiot, La Bruyère situates them on the side of the machine. Both represent linguistic aberrations that push them over the borderline of the human into the space of linguistic nonreferentiality and "gratuitous improvisation." In case of the courtier and the automaton, linguistic aberration is the sign of moral and intellectual failure: in the former, lies are the symptomatic production of the courtier's pathology; in the latter, it is nonsense. The moralist exercises the force of judgment as a kind of police action on the generalized inauthenticity of life at court: "When politeness becomes unreliable because 'refinement' suggests a possibility of corruption and loss of primitive veracity, one must rely instead on laws and social and political institutions as the basis of order (police)."[40] The moralist tries to restore linguistic authenticity in a degraded world order: this critical activity is both punitive and reactionary.

Giving the nonsense and the lie to the machine is an important moment for the genealogy of our morals: let us recall that in the cases of both the courtier/watch and the idiot/automaton, what occurs as moral censure takes place as assimilation of human being to machine, and the reduction of human activities to the satisfaction of bestial appetites. The aberrations of the empty word, the tautologies of the machine, are compared with and assimilated to a mechanics of idiotic repetition: this is, after all, what Bénichou criticizes in ambitious literary critics when he describes how he came to his own nonmethod of literary criticism. Because of his disappointment with Marxist doctrine on cultural production, Bénichou describes the itinerary of his skepticism:

> New systems that are more and more ambitious have created furor and inspired disciplines and hypotheses that are foreign to literature. Because these disorienting detours outside of the reality of the works have asserted themselves as methods, whoever did not adopt one of these above-mentioned methods, Marxist, psychoanalytic, or structuralist in one form or another, fell under the suspicion of—lamentably—not having one. In fact, one baptized as methods preconceived, systematic views on the real, which defined or assumed to define what literature *is:* a disguised projection of economy, the disavowed expression of unconscious drives, and an organization of verbal forms and signs, and etc. The methods of work that accompany the theory in each of these cases have not served to establish it: on the contrary, they are the result of the theory itself; it is an ensemble of procedures destined to confirm its own truth, extracted from a foreign source and imposed on literature as an a priori. In literary criticism, one should only call *method*, in the strict sense of the term—that is, a way of approaching a truth where nothing is presupposed about the nature of that truth—the approach that consists of gathering, handling correctly, interpreting in a plausible manner, a sufficient amount of information. That is, one should avoid all mental regions where that which is impossible to prove and that which is impossible to refute are one.[41]

Bénichou takes dictation directly from the historical periodicity itself. Who could contradict him without implicating herself in the implausible, the impossible, and the unreasonable? According to Bénichou, the critics who subscribe to a false method think that they know what literature is. They produce a false idol, that is, a literary object in their own image. Bénichou thinks that he has escaped from the specular trap when he grounds his literary object in the "expression" of the author's will. The magical communication that takes place between will and text is what grounds Bénichou's project.

In refusing this ground, other critics go wrong. Like courtiers and idiots, they turn in circles, whereas he is able to approach the truth. In fact, what we find at the end of this long citation is a final corrective measure that has to do with misnomers. These other theories have been misnamed. These theories are not methods—they are mechanical systems, producing prefabricated insights. Marxist, psychoanalytic, and structuralist theories are not true methods. They have tried to usurp the place of real literary criticism, and in so doing they have made a monster out of literature. The true method of literary criticism is Bénichou's nonmethod: here the path to truth is revealed as one that is not only self-effacing, but self-erasing. Others may get lost on ridiculous detours while chasing after chimeras. Like Descartes, Bénichou has, however, learned to proceed free of all prejudice and presuppositions toward the object of study. The denigration of theory as a demonically mechanical system becomes a textual motif that traverses a very different book—David Lehman's *Signs of the Times: Deconstruction and the Fall of Paul de Man*.[42] In Lehman's account of the case of Paul de Man, as in Bénichou's criticism of literary theory, the machine is hard at work, producing prejudices that appear self-evident.

As we have seen, the machine is a figure for the kind of knee-jerk, autonomic-nervous-system function that is the main feature of what we shall call for the moment "nonthinking." What thinking (in the academy) might entail is a reflection on the conditions, both historical and institutional, of the recent history of conflicts surrounding the place and legitimacy of literary theory. It will not be my argument that literary theory is equivalent to thinking. Literary theory has created enough of a conflict in literary studies, however, that we can look more closely at the rhetoric of this conflict. This is what I have tried to do by reading Paul Bénichou with Paul de Man.

Chapter 2
"What's the Difference?"

IN HIS ESSAY ON "THE UNCANNY," FREUD reads Olympia, the automaton of E. T. A. Hoffmann's story "The Sand-Man," as a double of the doomed protagonist Nathanael: she is the incarnation of his paralysis and impotence vis-à-vis his father.[1] The father, at least according to Freud's reading, is a two-faced figure, who on the one hand is a weak, loving figure who is eventually killed, and at the same time an evil patriarch who is murderously successful. Freud's theorization of the uncanny takes place through his reading of automaton in relationship to the two-faced father. These two faces are evoked by the two attitudes of Nathanael: one lovingly submissive, the other frozen and powerless. In psychoanalytic theory, machines and automatons are almost always doing the work of doubling: they are always ready to assume the metonymical burden of projections and displaced investments. In Freud's reading of Hoffmann's story, the automaton offers itself up as a particularly compelling figure of ambivalence. Literary ambivalence becomes readable to Freud in the unfolding of events around Nathanael's ill-fated affection for a machine.

Walter Benjamin demonstrates that the projection of the human gaze onto the inanimate object creates the uncanny force field known as aura; it is in this context that it becomes clear that Olympia's auratic power is related to the shiny lifelessness of her gaze. In seeing himself in the inanimate gaze, Nathanael experiences a shock of recognition. After buying a looking glass from Spalanzini, the peddler of optical devices, Nathanael trains it on Olympia:

> For the first time now he saw her exquisitely formed face. Only her eyes seemed particularly fixed and lifeless. But as he continued to look more

and more intently through the glass, it seemed as though moist moonbeams were beginning to shine in Olympia's eyes. It seemed as if the power of vision were only now starting to be kindled: her glances were inflamed with ever-increasing life.[2]

Nathanael sees his own passivity in Olympia's glassy eyes, but the more he looks at her, the more alive she becomes. The force of anthropomorphic projections makes it possible for objects to look back on us. Benjamin describes one of the conditions of the constitution of modern subjectivity in this tropic impulse. This shock of (mis)recognition is the sign of auratic decline: as experience becomes more and more impoverished, inanimate objects are the only things that hold our full attention. When Nathanael looks at Olympia, he, like the viewer,

> attains self-consciousness (that is, becomes a subject capable of perception) by being mediated through what it views. As de Man notes, such a mediation involves a negation since the perception of consciousness is derived from a discontinuity it cannot account for in terms of itself. What the object reflects is the viewer as a metaphor for the object's ability to view and reflect the viewer.[3]

The automaton/machine is fascinating precisely because it looks as affectually impoverished as the lover, at least in Freud's version of the story. Olympia brings Nathanael, however, back to life, and she stirs him out of his depression in a way that no one and nothing else can.

Machines can always be read as uncanny doubles of our own inanimation, but more important, for Benjamin the nature of experience itself under high capitalism is forged in the crucible of mass production, mass reproducibility, and homogenization. The rise of economic forces whose rationalization of the everyday has an increasingly insidious effect on everyday life, but the enslavement to the machine is an image that emerges for a generalized sense of powerlessness before invisible forces, paternalistic in nature, whose unrepresentability, except as an infernal principle of the machine, is only one figure of the complex nature of capital's power. In the study of literature, however, it becomes obvious that a need to denigrate or repudiate machines becomes one symptom of the way in which the questions of mechanical reproduction are repressed. For many critics, it remains a scandal to think of the text as a thing that works, and even more scandalous to think of a reading as participating in mass production. Denigration of the machine participates in the history of literary studies itself and designates one tendency to criticism that is resistant to a theorization of repetition:

> Traditional literary studies habitually use the language of machines in a negative way, deploring the mechanical and the technical as the death of values attached to life, form, inspiration, and so on. At best, a "technical" use of concepts is accorded an uneasy neutrality, without ever being allowed to become the heart of the matter. Machines *repeat,* and repetition means danger—compulsion and death.[4]

The most immediate threat embodied by the mechanical is repetition itself: in the denigration of the machine, literary criticism repeats itself, and in order to cast off the shadow of repetition it must all the more vehemently distance itself from the machine. It is in its repudiation of the forces of mechanical reproduction, and the experience of auratic impoverishment, that traditional literary studies participate in the decline of the literary object's aura.

Machine in both French and English has two highly divergent meanings. *Machina* is defined as "any artificial contrivance for performing work": this yields an instrumentalizing attitude; the second definition of *machina* as "device, contrivance, trick, stratagem" reveals a darker side of instrumentality itself. In Cassell's Latin dictionary *machinor* is defined as "to contrive, to invent, to devise, to plot evil." *Machiner* and *to machinate* are verbs that describe the hatching of plots and the conspiring of the agents of intrigue. Nathanael sees plots against him everywhere but is blinded by the lifeless gaze of the automaton. Meeting the anthropomorphic gaze of the automaton/machine produces a shock that is both difficult to account for and even more difficult to represent. In his reading of Hoffmann, Freud identifies the encounter as one that results in a paralyzing, disjunctive recognition. In anthropomorphizing the gaze of the machine, Nathanael falls in love. De Man taught that while anthropomorphism is not a trope, it allows for very precise kinds of reading, and a very specific form of love for literature, as an expansion of the idealization of human values, authenticity, and so on.[5] If he invites us to trade in the text-as-body for text-as-machine, he refuses to accept a certain description of literary criticism that is based on projecting a humanizing or even intellectual outlook that might emanate from any text. De Man attempts to confront a certain level of reading that "proceeds mechanically and unthinkingly."[6] In addition to de Man's reading of Kleist's *Über das Marionettentheater,* his refusal of a projective, humanizing gaze is articulated in his reading of

Kant, where the *Augenschein* is "destitute of all intervention of the intellect": "No mind is involved in the Kantian vision of ocean and heaven."[7] According to Rodolphe Gasché, de Man is describing "a purely material vision," and one whose contemplation of the aesthetic object yields a level of radically formal interpretation.[8] This refusal has far-reaching consequences in the debates over the constitution of the literary object, and the terms of literary interpretation, but it most immediately produces an affectual detachment that calls up deep suspicion. John Guillory criticizes de Man's eighteenth-century mechanical materialism, to use Fredric Jameson's terms, as being ideological determinism.[9] The inadequacy of de Man's account of the materiality of the signifier notwithstanding, an implicit critique of an anthropomorphizing and reductive phenomenology that Guillory does not take into account can be found throughout his work.[10] De Man's critique of anthropomorphism is not a simple one: it is predicated on a reading of the shock of recognition and decline of the aura described by Benjamin as both historical and political. In Benjamin's account of history, the shock punctuates experience as radical suspension and transforms a merely phenomenological account of the object into a historical and material one. De Man's refusal to give a human face to the act of reading or aesthetic contemplation calls up forms of resistance that are crucial in understanding the significance of the de Manian intervention in literary criticism: it takes place as an attempt to account for the radically particular and disjunctive qualities of the literary object.

In addition, if we are to take seriously Jameson's characterization of de Man as an Enlightenment materialist thinker, we must conclude that in order to understand the significance and limits of his work, we must examine the most symptomatic and virulent forms of resistance against his thinking. The work of resistance and refusal that I will examine closely here is David Lehman's *Signs of the Times: Deconstruction and the Fall of Paul de Man*, which offers itself up as the definitive assessment of an intellectual movement. In this text, we find the condemnation of de Man as both thinker and person. One of the crucial ways that Lehman does this involves an implicit denigration of the machine and the mechanical, and an explicit identification of de Man's and deconstruction's strategies of operation as a nefarious machination of soulless drones. If deconstructive critics are merely copies of one another, the idea is that other kinds of thinkers are singular, spontaneous, and authentic. The evil machine is opposed to the humanist once again, but like Nathanael, Lehman seems to be hypnotized by a destructive quality in the glassy-eyed object of his contemplation and hatred, which turns out to be an attitude in his work. In this, Lehman's own work crystallizes an antitheoretical position whose attempts to bury deconstruction coalesce in a parody of the tactics that he purports to deplore.

One of the most important criticisms of deconstruction is directed at what Lehman perceives to be its destructive intent: with a text in hand, the deconstructive critic behaves like Geraldo Rivera, riding roughshod over literature and philosophy in order to get the story of complicity and contradiction. Yet it is obvious that what is being destroyed in this book is any possibility of rational dialogue, any space of contention that might give us a way of coming to grips with deconstructive theory and its legacies. The intent to destroy is most remarkable precisely in Lehman's attitude vis-à-vis deconstruction as he sees it embodied in the work and person of Paul de Man. When Lehman writes about deconstruction's imperative to destroy what it reads, he is describing to us the way in which he is going to treat the case of de Man: "The critic must expose the text as one would expose a scam or a sham, for all texts are presumed guilty, complicitous with a Western philosophical tradition that the procedures of deconstruction are designed to discredit."[11] With talk-show-host enthusiasm, he himself performs the exposé of the exposé. Treating its objects of critique as guilty by association, and hell-bent on an agenda to discredit the texts it reads, deconstruction, at least in Lehman's caricatured portrait, demolishes literature with the weight of its own agenda of destruction.

Lehman even self-consciously indulges in readings that are based on a "deconstructive" playing with words. Rather than doing the work of reading, Lehman tries to convince us that the very "sound" of deconstruction should make us wary: "To the skeptical layman, as suspicious of jargon as the deconstructionist is suspicious of Western metaphysics, the sound of *deconstruction* suggests another possibility. Mightn't it be a *con* game concealing a destructive intent?" (41). The skeptical layman is Lehman himself, of course, and he reassures any reader who is disturbed by the difficulty of deconstructive criticism or theory. If Derrida is difficult to read, it is probably because he is just trying to pull one over on us. Lehman participates in the will to dismiss complexity or difficulty as a trick or a ruse, and his tyranny of common sense and simplicity devolves to a level of surprising viciousness. Practitioners of deconstruction have become utterly reprehensible. They have acquired the worst qualities of the courtier: they are abjectly ambitious and mechanical. Ambition and the machine fuel the very engine of their complicity. In describing the covert operations of deconstruction, Lehman is describing his own destructive strategies as a reader. When he writes that deconstructive critics presume the guilt of the texts at hand, he is describing his own attitude toward those critics. His own procedures are designed to discredit the fundamental project of deconstruction itself, this time in order to defend Western civilization from the zealots of theory, but the spirit and the mood of his book are anything but civilized. He accuses the acolytes of deconstruction of being nothing less than false prophets who serve the dark gods of ideological complicity.

He describes so-called deconstructionists as "a daisy-chain of brown-nosers" who are "unbelievably mechanical and wooden at the same time" (27). According to Lehman, deconstruction is a clearly defined network of like-minded professors who fiercely promote one another's work and use their institutional power to further the cause. Mastery of empty jargon certifies the budding theorist's professional standing; initiates are rewarded with teaching appointments and prestigious postdoctoral fellowships.[12] Deconstruction is the enemy within that eats away at the integrity of the humanities. Deconstructionists replicate themselves through a process of conversion and initiation. Like La Bruyère's courtier, the deconstructionist functions only with regard to ambition, with nothing more than (self-)promotion in mind. Instead of acting alone, however, deconstructionists are networked to one another in a conspiracy of "like-mindedness," exercising their authority in order to foster the growth of "budding theorists." In short, the disciples of theory are produced, like so many pod-people in a fantasy worthy of *The Invasion of the Body Snatchers*. They are robots who are programmed to not-think; they are engaged in the nonthinking furthering of their own careers.

Like the courtier, Lehman's theorists engage in mechanical rituals of propitiation. The theory robot is an impostor in the halls of academia, who, according to Lehman, has invaded the rank and file of humanists by attaching itself parasitically to intellectual institutions and speaking a highly coded jargon that passes for learning. Such robots are the very figures of dissimulation, and the jargon that they speak conceals not only corrosive emptiness but also "destructive intent." The skeptical Lehman is not, however, so easily duped. It is clear, from his description that the theory robot is a monstrous, complicitous, mechanical, and idiotic foil to the Lehman/Layman's heroic skepticism, for he is none other than a kind of Super-Layman, the courageous defender of literature, art, truth, and common sense.

The work of the theorists and their disciples is a travesty of scholarship, and yet they can "pass" and are even accepted unquestioningly in many walks of academia by those innocents unaware of their pernicious plan: "The gurus of deconstruction have been remarkably successful at recruiting disciples and turning them into promulgators of the faith" (72). And yes, they have a plan, they want to take over, and they follow orders like members of the Mob (recall the much-cited image of the Yale Mafia) as well as manifesting the worst aspects of a violent, mob mentality.[13] In return for the initiates' unquestioning submission, the cult leaders distribute favors to the faithful. Theory is a cult of pure destruction, a "terrorist sect." It is, however, a terrorism of opportunism and ambition. The leaders of the movement are not only authoritarian bullies—they are perverts and seducers: "To read Derrida at length, which is how he asks to be read, is to be expose oneself to a mesmerist's power. Immersed in Derrida's article "Biodegradables," one feels the full force of his fury—and one understands the seductive attractions of submitting to his rhetoric" (257). There is something irresistible in the power of a charlatan, a mystifier, a hypnotist, a seducer. While other scholars—real scholars—cultivate knowledge, educate students, write out of the spontaneous movement of real imagination in order to further the understanding and appreciation of real art and real literature, the theory robot is produced on the assembly line of abject opportunism, programmed to take over humanities departments and sow the seeds of discord and destruction in their path.

Yet the surprise about Lehman's dramatic account of the deconstructive conspiracy is its failure, and he is the one who will bear witness to that event. How did deconstruction come to lose the ground it had gained? Most important, the scandal of Paul de Man's wartime journalism broke while the other cause of deconstruction's demise was purely tactical. The scandal broke when it was revealed that in his early twenties during the German occupation of Belgium he wrote a series of articles for the Nazi-controlled Belgian daily *Le Soir*. In these articles on literature, he gave voice to a blatant anti-Semitism. De Man's wartime journalism was ground enough for Lehman to make wider insinuations about the hegemonic agenda of deconstruction in the humanities. But, according to Lehman, while deconstruction made great progress in its conspiracy to take over American academia, it did not know when to stop winning: "In the ten years we have gone from the image of a military phalanx crushing the resistance in its path to that of an overextended army whose supply lines are in trouble" (259). The analogy between the Third Reich's tactical errors and deconstruction's own militarized ambitions gone wrong is abundantly clear. In an earlier article, Lehman quotes an unnamed Ivy League professor gleefully declaring that "deconstruction turned out to be the thousand-year Reich that lasted 12 years."[14] As "proof" of deconstruction's decline, Lehman offers us a brief reading of a 1990 story by Christopher Tilghman called "In a Father's Place." Lehman's handling of the story offers us an example of his technique as a literary critic. The action takes place at the summerhouse of Dan, a widower and "the scion of a landed estate":

> Dan, a sympathetic fellow, feels estranged from his son Nick, an aspiring novelist. Nick has brought his friend Patty with him, and she is the villain of the piece. Contemptuous of Dan and his ancestral house, Patty spends most of the weekend in an antisocial posture: reading a book Tilghman identifies not by its title but by its author, Jacques Derrida. What does Nick see in Patty? She "tore the English Department at Columbia *apart*," Nick says. "I've never known anyone who takes less shit in her life." Patty wouldn't mind driving a wedge between Nick and his family. Nick, she says, is "trying to deconstruct this family" in his novel. "Deconstruct? You mean destroy?" Dan replies. The story reaches its climax when Dan throws Patty out of the house: "Oh, cut the crap about his work," he says, "you want his soul, you little Nazi, you want any soul you can get your hands on."[15]

For Lehman, this story is a sign of deconstruction's demise: a wealthy and sympathetic scion saves his son from a vicious, Derrida-reading woman who is also a social upstart, a guest who becomes an unwelcome intruder in the summerhouse. The villain is the woman, whose characterization as a soul-sucking harpy seems to warrant no further comment or analysis. There is the good guy, family man, our sympathetic Dan; the villain, Patty; poor Nick caught in between the two. Is the happy ending a surprise? It is enough for her to be an ungrateful interloper and a young woman with a bad attitude for Dan to call her a Nazi. This seems to endorse Lehman's position that deconstructive sympathies lead to right-wing fanaticism. But the ideological bias of Tilghman's story and Lehman's plot summary gives us another clue as to how Dan, Tilghman, and Lehman mobilize the drama of calling someone a Nazi in order to expel unwanted guests from the summer estates of sympathetic scions: it is a conservative agenda that this kind of accusation serves. *Nazi* is emptied of all historical meaning and designates here a theoretical dogmatism. Lehman offers the plot summary as if it should be self-evident which side we want to be on: readers are asked to endorse the idea that in "a father's place" we would have done the same. Misogyny and the hatred of deconstruction are entangled with one another in Tilghman's story and Lehman's plot summary: the threat to a paternal order—there is even the pretense of an aristocratic order ("wealthy scion of a landed estate")—has to be purged from the scene. That the threat is called a Nazi is supposed to leave Patty indefensible, for who would dare take her side after such a name has been evoked?

The fall of Paul de Man is described by Lehman as an allegory of everything wrong with deconstruction: de Man's moral and ethical compromises betray the moral and ethical compromises of deconstruction itself. As an example of de Man's personal dishonesty, Lehman cites the example of a letter to Renato Poggioli, director of the Harvard Society of Fellows, in which de Man seeks to defend himself not only against accusations of collaboration under Nazi occupation, but also rumors of dishonest business practices having to do with the bankrupt Editions Hermès.[16] In the letter, de Man claims that Hendrik de Man is his father. Hendrik de Man was a psychoanalytically oriented socialist, a prominent figure in Belgian politics who advised King Leopold to cooperate with the Nazis.[17] Lehman interprets this claim as a sign of de Man's "cunning." In the letter to Poggioli, dated January 25, 1955, when an anonymous denunciation of his activities in wartime Belgium arose, he merely testifies to having written for *Le Soir* during the first years of Nazi occupation in Belgium. Lehman decides that de Man lied to Poggioli for the following reason: if the controversial Hendrik de Man were his father, Paul could say that he was merely following in his father's footsteps when he decided to write for the Nazi-controlled daily. According to Lehman, Paul lied in order to protect himself prophylactically from future accusations of complicity. If Hendrik were his father, however, Paul could claim in self-defense and under pressure that he was merely trying to be a good son.[18] In Lehman's version, Paul de Man's revisionist genealogy is the sure sign of his desire to stage a cover-up.

In Richard Klein's "The Blindness of Hyperboles: The Ellipses of Insight," he refers to Hendrik de Man as Paul de Man's father.[19] Klein's attempt to read in de Man's psychobiography the signs of an ambivalent repression of psychoanalytic insight—first and foremost in the nondialectical description of blindness and insight and later in the discussion of literary history—aims at criticizing de Man's work while at the same time paying him tribute. In the mode of critique, Klein attributes de Man's fleeing from and returning to psychoanalysis as a symptom of his relationship to his *father,* Hendrik, who, according to Klein, was one of the "first serious thinkers to apply explicitly Freudian categories to the analysis of alienation" (42). Klein cites de Man's correction of his genealogy: de Man writes to Klein to let him know that Hendrik is not his father, but that he is nevertheless wary of psychoanalysis. When de Man writes to Klein, "My skepticism doesn't spring from the fact that Henri [sic] de Man is my uncle and not my father," he does seem to leave room for the fact that this skepticism is in some way related to Hendrik (or Henri). Klein goes on to insist on the importance of avuncularity and alludes elliptically to the status of parental siblings, ending his article with a rhetorical question, "What, after all, is an uncle?" (44). Although this question keeps open the problem of a psychoanalytic intervention in the analysis of de Man's resistance to psychoanalytic theory, Klein ensures himself against paternal censure or wrath by praising his two theoretical fathers, Paul de Man and Jacques Derrida, leaving his readers with the idea that despite certain blindnesses on the part of the latter, the two figures supplement each other in such a way that we can only conclude that "Fathers know best":

> In conclusion I will add that the possibility of this article, its stance and its rhetoric, is no doubt determined by my own relation to de Man and Derrida, both of whom are my teachers. The good fortune of having double fathers is enhanced by the fact that . . . each father's text has been deeply engaged in reading the others. . . . And like the best fathers, they do not efface themselves, nor do they take their roles literally. . . . In other words, they give themselves sons who can acknowledge their existence as fathers as a necessary fiction. (Klein, 43)

This touching and embarrassing display of filial piety is an excellent example of the kind of idealization among de Man's students, which John Guillory will later identify as symptomatic of the kind of transferential fixation that de Man's teaching and personality produced. In Klein's article, we see the attempts to secure a familial bond with his teachers, especially de Man: the son's ambivalence, however, is only readable as a contradiction in his own text. Although Klein claims that one of de Man's qualities as a father has to do with not effacing himself as such, earlier he praises de Man for his great powers of self-effacement: "There are very few modern critics . . . who are able to efface themselves so totally in front of a text" (33). Klein attempts to break through de Manian impersonality by expressing his love for his father/master by bringing psychoanalytic theory into the picture. What Klein praises in de Man is exactly what the son is unable to perform: he cannot efface his own desire for a father before de Man's text. Klein's essay would support Guillory's reading of de Man's refusal to acknowledge transferential aberrations as highly symptomatic and very damaging in the long run to the intellectual development of his students because the denial is absolute. According to Guillory, this denial promotes a kind of neurotic fixation: "Discipleship is an inescapable pathology, then, but not like the 'artificial neurosis' of the transference. The transference in pedagogy is least likely to be dissolved when the thought is transmitted in the 'impersonal' form of a science."[20] Klein claims that de Man does not efface himself as a father, but this, too, seems to be a phantasmatic formulation. The fact that de Man gives himself sons who are capable of recognizing his fictional paternity is based on Klein's fantasy of a father who would destroy himself so as to spare his son the unpleasant task. The son, then, like Hoffmann's Nathanael, can remain in a position of passive adulation before the best of fathers. (This is much harder for daughters, but we will not take up that problem now.)

An uncle might be like a father insofar as he represents another aspect of one's paternity; in hindsight, the figure of the uncle brings together psychoanalysis and complicity with Nazism. Paul as son (or nephew) reacts negatively to his avuncular and paternal legacy by avoiding psychoanalytic insight, until, according to Klein, he rediscovers Freud again, through the work of Jacques Derrida: "He can only encounter Freud, read the Father, in the guise of Derrida, in the position of a younger brother towards whom de Man can play the tutor" (Klein, 43). Hendrik is also the name of Paul's older brother, whose premature death deals a fatal blow to Paul's mother, and "real" or biological father, Robert. There were two deaths in the family that made Robert de Man give up the care of Paul to Uncle Hendrik—the death of Paul's older brother and the suicide of his wife a year later. By giving Paul to his brother to raise, the father renounces his paternity in a radical and violent gesture. Paul is in a sense disowned, but not completely. The deaths are apparently too much for Robert, who instead of mourning his losses decides to join the dead by giving up his surviving son. The father joins his wife and older son by effacing himself as father of the survivor. In the context of a father who writes himself out of the picture, Paul de Man's own self-effacement as a critic (which Klein returns to again and again in his article) turns out to have a denser signification. Self-effacement runs in the family.

What is the young Paul de Man left with, but his dead brother's namesake as guardian and father-figure? The appearance of an avuncular inheritance confirms Laurence Rickels's theorization of mourning and metonymical displacement: parental siblings do double duty as the targets of ambivalence and the placeholders of unbearable losses.[21] Paul loses a brother and finds him replaced by his namesake, the original Hendrik: one Hendrik is exchanged for another. That Paul would later name Hendrik as his father is a gesture of duplicity and aggression, but it also articulates a truth. Both affectively and intellectually, Paul names the man to whom his father ceded his place. The father, Robert, never reemerges in the biographical accounts, except as a ghost and palimpsestic sign of the son's duplicity and guilt. Paul has to stick it out with his psychoanalytically oriented and powerful uncle.

Klein is cautious about psychobiographically reducing de Man to an interesting case history (a criticism that de Man reserved for bad readers of Rousseau). In other words, Klein remains a faithful disciple. It is possible, however, to argue for a stronger psychoanalytic analysis of Paul de Man along the lines of the theorization of writing and repression, forgetting and deferral, that structures the work of mourning and the accounting of losses, which are extremely complicated to sort through, especially after the last World War. If one does reduce Paul de Man to a case, and a case of history, it is in order to continue the work that Klein began, to look for the traces left there by the body count of history, and the force of psychoanalytic theory deferred.[22] The guilt in Paul de Man's psychobiography begins before the war, in his having already survived two deaths; the drama of his abandonment by his father is repeated, the second time as renunciation, in order that his own guilt be renewed in a lie. Klein's article was written well before the discovery of de Man's wartime journalism: it is a testimony to a certain kind of blindness and insight into his own transferential double binds. The way in which Klein remains attached to a certain idealized image of the objects of his admiration appears in hindsight as error, but a contingent one, whose consequences he, like all of de Man's students, would have to suffer.

Returning to Lehman, he sees in the demise of deconstruction a resolution of the crisis in the humanities, which for him comes from abroad as a kind of moral ambiguity from which all of us (in the United States) are free. He fails, however, to bury deconstruction with de Man's guilt: in declaring deconstruction's defeat, he only proclaims himself to be on the side of the winners, those who would preserve the integrity of a certain mode of reading for literary genius and artistic greatness. By the end of his book, Lehman has successfully made *destruction* a synonym for *deconstruction:* that he in turn seeks to destroy completely the work and reputations of people like Derrida and de Man is part of his attempt to preserve once and for all his right to dictate the terms of the debate. The excessive devotion of de Manians does at times make them uncanny, but Lehman's desire to destroy the object of their worship is the other side of that infantile, unreasonable, and idealizing infatuation, and it is called hate.

In constructing the association between deconstruction and National Socialism itself, Lehman tries to protect literary criticism as he knows it from its most serious challenger, and he tries to draw the debate about literary theory to a close. The fact that his work has been influential in journalistic circles is evident in the work of Michiko Kakutani, cultural critic of the *New York Times,* who participates in the characterization of deconstruction as a defeated ideological fanaticism with extreme right-wing tendencies. Such insinuations have become symptoms of the ways in which contemporary conflicts and disputes are mediated by dramatic and damning associations. Lehman's criticism marks an extreme of antitheoretical venom, but it is not at all uncommon to find his views echoed in many different areas of popular discourse about deconstruction. As much as Lehman wanted it, however, this book did not write the demise of deconstruction; what it made almost impossible was serious criticism of the work of Paul de Man. As a document of one side of a conflict, it is testimony to how debased intellectual debate around the question of Paul de Man's wartime journalism had become.[23] The book also lowered the standard of public discussions of intellectual conflicts in general. Paul de Man's ambivalent legacy should be read and thought through in the context of resistance to psychoanalysis. His case is a case history, but rather than letting psychoanalytic theory win this war, de Man's own insights and blindnesses have put pressure on the ways in which we think about the recent history of theory in general.

The machine has been framed in order to establish a ground on which the human being can be represented. Lehman's "mechanical and wooden" theorists allow him to defend the precincts of the spontaneous and the biological. This prejudicial setup can be read, thought through, and theorized as a temporal aberration: machines keep time through thoughtless repetition (this is what theorists, courtiers, and automatons do); the human and the organic live in a temporality free from the constraints of infernal repetition. We may depend on the machine to measure time, but our time is vastly different from the empty minutes of La Bruyère's courtier/automaton. In the critique of theory, we find that the machine becomes a figure for the infernal ambition of the subject of tyranny and in this way acts as a foil to a mode of heroic humanism. Benjamin's critique of historicism is based on his suspicion about certain modes of secular timekeeping. Both Benjamin and de Man are concerned with the problem of history and historicism as a condition of anthropomorphization: the tropological question is intimately related to an aesthetic ideology of temporality that is implied in every act of reading and every gesture of interpretation. Anthropomorphized time allows for periodization in literary studies, but it also produces a haunted and doubled body of literature. Theoretical study of literature and literary objects brings about the other aberration that David Lehman finds so deplorable: literature departments are suddenly open to studying film and other forms of contemporary popular culture. Differentiation is the question here, and the de Manian analysis of the question "What's the difference?" (between film and literature, theorist and humanist, automaton and human being) offers up a challenge to any traditional search for meaning in and outside of literature.

The film *Blade Runner* allows us to frame the de Manian question by allegorizing the representational apparatus itself. *Blade Runner* is the film adaptation of Phillip K. Dick's science fiction novel *Do Androids Dream of Electric Sheep?* In Dick's novel, Deckard is an unhappily married, struggling policeman, unable to move off world, that is, off pollution-ravaged Earth to one of the planetary colonies. He is engaged in the extermination of renegade "replicants," or sophisticated androids created to perform work that human beings no longer want to perform. The replicant is also in a sense cinema, itself always already a double of the dream work and dependent on the principle of projection. What is of interest for me is the way in which texts, cinematic and literary alike, may be read as machine-like, producing and productive of meaning in an automatic way—automatic insofar as the machine of representation can always function autonomously, independent from the intentions or the psychology of its creator/author. Every text is a bit of a Frankenstein, and every writer is distinctly marginal to his/her text. The machine can be read as an allegory for inefficacy of intentionality when it comes to the question of writing. The misfiring of any original intentionality can be considered a side effect of a paranoid mechanism that distorts, deforms, defers, and disguises desire. The distortion, deformation, deferral, and disguise of desire take place as a function of the specificity and peculiarities of writing (as representation of psyche). In Derrida's reading of Freud, we can see how the machine allegorizes a psychic apparatus that is always on the defensive—on the verge, if not in the process, of approaching paranoia.

"What's the Difference?"

Representation is always about a process of transformation and substitution: it is tricky, relying on contrivances and stratagems *(machina, machinor)*. For de Man, the problem of representation is always a problem of reading: "How is reading represented in writing?" What is the distance between the representation of reading and the reading itself? This distance is both difference and *différance,* or the spatiotemporal deferment of allegory. The time of reading replicates in a compressed and accelerated form the duration of writing: the run time of the film compresses the production time as well. The differential and untimely aspect of the time of representation and its effects is what Derrida refers to when he reminds us that *Nachträglichkeit* refers to both belatedness and supplementarity. The cinematic present is reconstituted and reproduced in the necessary rupture between time of production and time of projection. Thus time of re-presentation is radically heterogeneous to the time of the reader, the time of the spectator.

"Film represents the double by being the double of the dream work," writes Laurence Rickels.[24] And so filmic space doubles the space of experience. In Ridley Scott's film *Blade Runner,* doubles, or "replicants," are such perfect cybernetic twins of human beings that it is *almost* impossible to discern the difference between human original and technological copy. The replicant contains within itself the principle of its own movement. Science has mastered the principle of the production of humanoid cybernetic creatures—it is automaton. The double has always been important for psychoanalysis, but "important" is perhaps an understatement. Doubling is constitutive of the relationship between thinking and mental illness. For Freud, the philosopher is the normative counterpart or socially acceptable double of the paranoiac, who is always hard at work constructing elaborate machines of interpretation.[25] The paranoiac's machine allows him to defer forever the formulation of homosexual desire. The paranoiac sees projected everywhere the exciting story of his never-ending persecution. Freud does not accord the normative counterpart to the paranoiac (the philosopher) many privileges in regard to his homologue in sickness. The paranoiac may be pathological, but it is in the structures of his pathology that we find an analogy to the philosopher's work. The philosopher also constructs machines that have to do with deferral: this machine is related to the posthumous potential of all writing.

The writing machine is on the side of death, absence, and the uncanny autonomy of representation as replication. Derrida compares writing and marking to machines: "To write is to produce a mark that will constitute a kind of machine that is in turn productive, that my future disappearance in principal will not prevent from functioning and from yielding, and yielding itself to, reading a rewriting."[26] Writing and leaving traces are both about a kind of radical dispossession: "I" the author give up the marks "I" have made to a future from which I will have disappeared. *Writing* here refers to my eventual failure to be present: writing, like a machine, can always survive its author/maker.

In "Freud and the Scene of Writing," Derrida points to the contradictions in Freud's treatment of machines when he emphasizes that in the *Interpretation of Dreams* (1900) Freud slips into a hermeneutic strategy that he himself has criticized on the part of others who have tried to analyze dreams. Freud refuses a certain kind of interpretation that treats dreams as so many romans à clef, and he is careful to point out the futility of such modes of translation. It is particularly the one-to-one correspondence of symbol to meaning that is questioned. However, as Derrida points out, "It will be said: and yet Freud translates all the time."[27] More precisely, in the *Interpretation of Dreams* the machine is "translated" into genitals: "It is highly probable that all complicated machinery and apparatus occurring in dreams stand for the genitals (and as a rule male ones) [1919]—in describing which dream symbolism is as indefatigable as the 'joke-work'" (*SE* 5:356). Dream symbolism is "indefatigable" in its descriptions of the genitals; that is, in some sense dream symbolism works like a translation machine, tirelessly transforming genitals into "symbols." Is Freud "translating" machines into genitals, and if so, what are the consequences of such a translation? Machines seem to be always *standing in* for something else insofar as they never stand alone. As Derrida has shown, Freud has recourse to the (writing) machine in order to represent the processes of the psychic apparatus itself. The "standing in" of machines for genitals is yet another machine: dream symbolism works like a machine, diligently turning genitals into hats, ties, and increasingly complex machines—stand-ins or substitutes that circulate freely in and through dreams.

According to Victor Tausk, machines in dreams "refer to the dreamer's own genitalia." But for Tausk, the machine becomes increasingly complex in order to rouse the dreamer's intellectual interest while inhibiting his libidinal instinct.[28] Machine dreams inhibit masturbation by deferring the possibility of intellectual understanding: the dreamer's intellectual interest is aroused while her sexual arousal diminishes. Genitals and machines exist in a relationship of metonymy because of *manipulation:* sublimation can take place because of the efficacy of metonymical substitution. Interest in genitals can become transformed into interest in machines. The machine sets into motion a series of displacements that slide from representation to representability as it never seems to be able to resolve its own "proper" meaning. Its ability to blossom into configurations of greater and greater complexity allows it to defer any kind of stability that can be easily and immediately "grasped" or "seized": its metamorphic and metaphoric potential arouses the dreamer's intellectual interest while displacing/deferring her libidinal instinct. As the genitals become machine-like, sexual manipulation becomes displaced or replaced by intellectual manipulation. To say that machines are nothing more than genitals reduces our analysis to a naive quest for one-to-one correspondences between signs and referents. We have to remember that the dream machine is always changing—that is, it is incessantly increasing in complexity: it takes on, in a manner of speaking, a life of its own. Both writing and masturbation are two activities that have to do with manual manipulation: both are jobs that demand a hands-on approach that critics like Bénichou deplores.

In returning to Descartes's *Meditations* by way of a reading of *Blade Runner*, I am following an itinerary of reading already drawn up by Slavoj Žižek and Kaja Silverman. (The film's protagonist is named Deckard, an obvious Anglicization of Descartes.) This science fiction film allegorizes the ambivalence of technological innovation, especially when it concerns the question of reproduction versus replication/mass production. The problem of the replicant, that beautiful and dangerous double of the human being, offers itself up for analysis as "the science fiction impulse" described by Rickels.[29] The European automatons fabricated during the seventeenth and eighteenth centuries inspire the Faustian science fiction dramas that haunt the history of technology's progress. It was the invention of the spring and spring-action mechanisms that allowed the craftsmen of the seventeenth century to create machines (or gadgets) whose source of movement was assiduously hidden from view. The gilded surface of a clock or watch could hide a complex, self-moving mechanism. The radical noncommunication of Baroque interiors and exteriors has something to do with the Baroque clock, whose mechanisms were completely disguised by its ornate container.[30] In painting, the calculation of perspective facilitated an illusion of depth and three-dimensionality of the picture plane; this technical and mathematical innovation allowed scientific progress to have very concrete aesthetic effects. An engineer and architect like Salomon de Caus who wrote on the question of perspective was also interested in automatons, solar clocks, and other fantastic devices like hydraulic machines; for him, all these objects played a role in creating wonderful illusions.[31] These objects were related to technical innovations in drawing, like foreshortening, that created deceptive visual effects of depth and three-dimensionality on the picture plane. Jurgis Balustraitis speculates that the work of Salomon de Caus influenced the way in which Descartes conceived of the functioning of the human body. De Caus constructed mythological grottoes and fountains in which figures were animated by a combination of hydraulics and mechanics:

> In taking his inspiration from these moving machines in his meditations on the structure and function of living organisms, Descartes moves outside the realm of logic into that of imagination; he thinks of the world as a theatre in which the secrets of nature are revealed through the medium of toys constructed by men.[32]

Descartes, however, overturns the dependency of thinking and perceptive cognition on the senses precisely because he was aware of the inherent dangers of this celebration of imagination, illusion, artifice, and theatricality. In *Technics and Civilization,* Lewis Mumford describes the seventeenth century as the climax of what he calls the eotechnic phase (1000–1750) in the history of Western technics. The eotechnic phase saw an acceleration of technological progress and brought together "dispersed advances and suggestions of other civilizations" while the processes of "invention and 'experimental adaptation' slowly accelerated."[33] It is perhaps this difference in the rate of technological progress that makes Descartes's attitude toward illusion and spectacle so radically different from Salomon de Caus's affirmation of the technical reproduction of human movement. Descartes became more and more interested in the philosophical problem posed by illusion: reason must be made the master of imagination.

The replicants of *Blade Runner* pose the question of illusion and temporality: they are like advanced cybernetic clocks that count down to their own deaths, but the technological principle of their movement is so well hidden that it is almost impossible to differentiate them from human beings. New "tests" must constantly be developed in order to secure the boundaries separating the human from its replicant. The question of difference remains fundamentally unanswerable for the man assigned to destroy them, Deckard/Descartes. "What's the difference (between a replicant and a human being)?" This difference cannot be apprehended immediately: it must be calculated. Deckard depends on a series of technological mechanisms to help him sort out the mechanical replicas from their human originals. The similarity of his task to the Cartesian project is far from coincidental:

> Descartes's effort to address the perceptual domain coincides with the attempts to redress its limits: he corrects errors and expands its scope so that it no longer has anything to do with visual perception, in a human rather than a purely technical sense. The development of instruments that expand the scope of vision coincides with the instrumentalization of visual perception as a whole, since technology enables a greater and more perfect knowledge of nature. Descartes connects the perfectibility of human knowledge with the extension of human perception through artificial organs. They expand the horizon of perception while erasing its experiential, all-too-human character.[34]

That these instruments or artificial organs could represent the very limits of human knowledge is the problem posed by the film: the replicants, the Nexus 6, travel places no human being could go; they have seen things that no human being has seen. They have become conscious of the fact that they have expanded the human field of perception, intervention, and exploitation. Descartes anticipates this effacement of human perception at the horizon of scientific progress. The replicants are Cartesian insofar as their difference is visually inaccessible: their hyperbolic resemblance to the original human beings whose shape they have assumed demands that other instruments for calculating difference be perfected. They represent the rebellion of pure instrumentality. Science must then perfect an apparatus for telling the difference in order to stabilize the boundaries between these almost perfect technological doubles and their human counterparts. For Deckard, however, the instruments of his investigation begin to fail him on the question, "What is the difference?"

Such a question, de Man reminds us, can be grammatical and rhetorical at the same time. It is grammatical insofar as a possible answer is implied by its very formulation: the answer would be, Nexus 6 replicants are manufactured by Tyrell Corporation, have a limited life span, have different retinal reactions from those of human beings under duress, are sociopathically murderous, are physically stronger and more intelligent, and so on. The question is a rhetorical one insofar as it invites the following response: "No difference at all," or else, "Despite the differences identified in response to the grammatical question, there is still no significant difference at all." What we have in this second, more elaborated answer is an implied negation of the question itself. Paul de Man has shown that the rhetorical modality has something to do with the literary realm of implicit rather than explicit meanings.[35] What drives the narrative of *Blade Runner* is precisely the implicit impossibility of "knowing" what the difference between replicant and human being might be, and the disquieting uncertainty surrounding the question of whether "knowing" the difference would make any difference at all.

The nonmimetic order of vision that Descartes develops establishes a phenomenal relationship to things that exceeds the limits of immediate resemblances. As Dahlia Judovitz puts it,

> The effort to establish a theory of perception—which on the one hand designates the object (the thing itself), and designates on the other hand its mirror reflection as another kind of a thing . . . while binding them together by the same mechanical causality—is to create an empirical world, in the modern sense of the word.[36]

In a purely empirical world where experience and sensation are held in the greatest suspicion, the threat of hyperbolic doubt must be contained. The imposition of doubt takes place as a gesture of violence. Deckard's investigations are not philosophical; they are implicated in the violent enforcement of the law. As an ambivalent noirish cop from the future, he doubles the seventeenth-century philosopher: he can be considered a replication and a revision of Descartes himself. Deckard, science fiction hero, doubles film noir protagonists at the same time that he doubles Descartes.[37] Deckard's dilemma is epistemological, ethical, and philosophical. He is a failed Cartesian because he is never sure of his hand: he and Descartes may have similar misgivings that arise from the question of "What's the difference (between a human being and the mechanical double of a human)?" but Deckard's question finds no sure answer.

As he engages upon the unsavory job of retiring replicants, the film implies an ironic rewriting of Descartes's conclusions reached during the Second Meditation. Descartes comments on the form of a piece of beeswax, recently removed from the hive: "[It] has not yet lost the sweetness of the honey that it once contained: there is still something of the odor of the flowers from which it was produced; its color, its shape, its size are apparent; it is hard, it is cold to the touch, and if you strike it, it produces a sound.[38] After the changes that the piece of beeswax undergoes when heated (it expands, grows soft, loses its floral odor, liquefies), Descartes still recognizes it as the same piece of wax because "he does not identify it by means of the senses, but by means of mind." He then relates how easy it is to fall into error, to be deceived:

> Words nevertheless stop me, and I am almost deceived by the terms of ordinary language; we say that we see the same piece of beeswax when it is presented to us instead of saying that we judge that it is the same piece of beeswax because it has the same color and shape; I would almost like to conclude from this formulation that we recognize the beeswax by one's vision and not solely by the examination of one's mind. If I should by accident look through the window at men passing on the street, at the view of which, I do not fail to say, I see men, just so I say I see the piece of beeswax. However, what do I see from this window if not hats and coats, which could very well be concealing ghosts *[spectres]* or simulated men *[des hommes feints]* who move only by means of springs? But I judge that these are real men, and thus I understand it by the sole power of judgment that resides in my mind, that which I believed I saw with my eyes.[39]

First, Descartes demonstrates that when we identify the piece of wax, we do so not by vision of one's eyes but, rather, by the inspection of the mind *(esprit)*. He might have been led astray by the terms of common language itself (we readily say, "I see the beeswax" rather than the more precise Cartesian formulation, "it is only through my intellectual inspection that I judge what I see to be the piece of beeswax"). But he is saved from such an error because he happens to look out the window and begins to embark upon another train of thought. At first upon looking out the window, he may be tempted to say that he sees men in their hats and coats, passing on the street, but he corrects himself. What is to say that under these hats and coats are not ghosts or "artificial men" who move by spring action, who are ingeniously designed automatons in fact disguised as men? It is the power of judgment that resides in his mind/*esprit* and not in his senses that allows Descartes to affirm that under their hats and coats, the creatures he sees from his window are not fake men but real ones. In order to arrive at the certainty of his judgment, he must pass through the moment of "hyperbolic doubt" ("What is to say that underneath these hats and coats, the creatures that I believe to be men are not ghosts or simulated men?"). We can imagine that there might be other circumstances under which Descartes could not so quickly dismiss the doubts that can arise when one is presented with purely sensorial information. He could reason in the following manner: "I see what looks like men dressed in hats and coats passing under my window, but I cannot be sure to judge them as such, because I know that there is a skilled automaton maker nearby who sends his automatons down the street at this time of day." In the absence of such conditions that would reinforce Descartes's doubt, he can affirm and judge intellectually that what look like men are to be identified as men.

Deckard's dilemma has to do with no longer being able to judge the difference between human beings and their replicas either by his senses or by his intellect. The difference is evaluated by "scientific" means, by the Voigt Kampff test, which measures retinal activity and reaction time to a series of questions. The test must be painstakingly administered to each subject, but even with the empirical results provided by such a test, Deckard remains uncertain as to the difference between the replicant and the human being; he continues to doubt. *To doubt* implies a vacillation between two places and a division between two poles. In Latin, *dubitare* is formed by *duo* and *habitare*.[40] To be unable to resolve one's doubts, then, is to be condemned to the exile of uncertainty, for she or he who occupies the space between two places "inhabits" liminality and lives in the margins of habitat itself. This is where Deckard lives, his address and his destination, at the same time.

For Descartes looking out his window, the difference between the simulated man and the real man is taken for granted: the mind identifies and judges. For Deckard, however, the scientific or empirical evidence gathered from the test that aims to guarantee that difference between the replica of the human and the real human being offers him little security. His distaste for the job of "retiring" renegade replicants is never overcome. Deckard's pathological and painful doubt could be formulated in terms of the famous Freudian articulation of perversion: "I know that [they are replicants] . . . but nevertheless I still believe [that they might be no different from human beings]. . . ." This formula has of course been used to describe the position of the filmic spectator and can be extended to address the problem of the reader of any fictional text. His uncertainty is established stylistically, that is, implicitly, in the melancholic atmosphere of a futuristic Los Angeles that never sees a sunny day, in his own loneliness and isolation.[41] His complicity with the violence of the police chief's dismissal of the replicants alludes to the weary impotence of every noir protagonist caught in a web of corruption, the limits of which he cannot conceive and which forces him into a choice between compromise/survival and resistance/destruction.

Blade Runner rewrites Descartes's question for Deckard in the following way: "How am I to know that underneath the hats and cloaks of the replicants I am 'retiring' they are not really human and not all that different from myself?" He cannot trust his sensorium, because he can see no significant difference between replicant and human being. The gaze of the filmic spectator is implicated here, for we, like Deckard, can see no difference at all. The status of how seeing might lead to "knowing" the difference is made altogether problematic. The spectator might "know" from information provided by the diegesis that a character is not human, but without such a sign from the narrative we would never be able to "tell" the difference. The replicants "look" not at all different from his or her human counterparts. What the film establishes when dealing with uncanny doubles of the human is precisely the unstable fiction of *difference.* For Dahlia Judovitz, the desire of the automaton to usurp the place of its human original allegorizes the threatening aspects of Cartesian reason in relationship to the field of human perception. Despite the obsolescence of the perceptual world, the violence of its reduction and abstraction produces haunting special effects:

> The Renaissance and baroque paradigms that problematized the question of representation (the world as theater) are displaced by the emergence of a concept of truth that elides its own representational character. Henceforth, it is the sensible world which takes on a spectral quality while reason aspires to a reality free of both artifice and illusion. Human invention reduces illusion to technical artifice, transforming nature into a mechanical phantom. Like an automaton that supplants its human model through a haunting resemblance, so does the Cartesian fiction of rationality threaten to render obsolete the perceptual world.[42]

Deckard distrusts empirical evidence (the test results) as well as the word of the law (the police chief's callous dismissal of the replicants as "skin-jobs"). For him, the world of easy certainties is increasingly *uninhabitable:* it is at this moment that he falls in love with a woman whom he has identified as nonhuman. The difference between "retirement" and "execution" pivots on the problem of self-identification, which the film never resolves: have the replicants in some way acceded to a certain kind of subjectivity that would render them "human"? Have the difference and the distance between technological replication and organic reproduction been reduced to insignificance?

If the question "What is the difference (between human being and replicant)?" is rendered invalid, Deckard would become a killer. The uncertainty about the other (the replicants) leads in the end to a terrifying doubt about the self (Deckard): "Am I or am I not a murderer?" In the original story, Phillip K. Dick's science fiction classic *Do Androids Dream of Electric Sheep?* (1969), there is a much more intense preoccupation with "real" animals and mechanical animals. Deckard is situated firmly in a more explicitly Cartesian schema as the rational and paranoid human being, calculating his difference from animals, machines, and animal-machines. For Descartes, the mechanical double of the human would never be able to use language in a spontaneous fashion: it could perhaps be "taught" or built to say certain phrases, but it would be nothing more than a parrot that has learned to repeat a few sentences. It would be deprived, then, of both language and what we might want to call linguistic intentionality. As we saw in the previous chapter, La Bruyère uses the notion of linguistic spontaneity in his descriptions of moral and intellectual failure.[43] *Blade Runner*'s replicants, however, are not only capable of using language: they are capable of using irony by playing with intentionality. When Pris, a renegade replicant, meets J. F. Sebastian, the eccentric recluse who designs cyborg eyes for the Tyrell Corporation, he asks her to "do something." Her response is entirely ironic: "I think, therefore I am(?)" she seems to ask impishly, quoting Descartes and commenting not a little bit sadistically on the absurdity of such an imperative. She knows full well how to obey and not to obey at the same time. Yet this does not do anything to secure her ontological credentials: her ironic rewriting of the formula of Cartesian existence has to do with the addition of a question mark. Cogito is henceforth marked as a *question*: the ambiguity of Pris's reply, "I think, therefore I am(?)" rests on its almost being a question.

Rachel, the replicant with whom Deckard has fallen in love, has been implanted with "false" memories and therefore has no knowledge of her "true identity" until she takes the Voigt Kampff test. Rachel is the assistant to Tyrell, who heads the corporation. It is at Tyrell's suggestion that Deckard puts the test to the test—that Rachel submits to it with resignation. Before taking the test, she is certain that she is a human being. After giving her the test, Deckard does not declare that she has failed to pass it; she divines it from his malaise. After the results are in, Deckard asks Tyrell, "How can it not know what it is?" Implanted memories are the answer to that question, but as Elissa Marder and Kaja Silverman have both shown in their work on *Blade Runner,* all memories are, psychoanalytically speaking, implanted. For Silverman, because of their implanted memories, the replicants become "hyperbolically" human; they "dramatize the fundamental terms and conditions of subjectivity." Insofar as they are "copies" of the human characters in the film, however, they are copies of copies: Deckard doubles for Descartes and film noir protagonist, and Tyrell is a copy of and reference to the evil genius, and so on.[44] In the film, no one passes the test of being human: every test subject fails. The only way to "pass" as human is to avoid putting one's humanity to the test.

Silverman's and Marder's work on the film affirms and confirms the increasing confusion of the organic original with its technological doubles. Silverman argues that "implanted" memories are a distinctly *human* trait and bases her assertion on the Freudian notion of *Nachträglichkeit, après coup*, or deferral,[45] as it is described in "Project for a Scientific Psychology."[46] Silverman asserts that the replicants[47] dramatize "the fundamental terms and conditions of subjectivity." According to Silverman, the replicants also "provide us with a paradigm" for the nonoriginary aspects of human subjectivity, and she ends her article with a reference to the replicants as being "all-too-human."[48] We could turn this formulation around and argue that the structure of human memory is "technological—all-too-technological." Reproducing memories is the work of the film, whose structure is uniquely hospitable to the formation of flashbacks.

Supplementarity becomes the center of the debate between Derrida and Foucault in their readings of Descartes. In Derrida's "Cogito and the History of Madness" there is a critique of Foucault's reading of Descartes's *Méditations*.[49] In *Madness and Civilization,* Foucault defines the Cartesian arrival at reason and certitude as one that is based on the exile and the exclusion of madness.[50] If Derrida criticizes Foucault's reading of Descartes, he is criticizing the way in which Foucault neglects the differential relationship between cogito and madness: the difference here is one of degree, of deferral. When Descartes describes the process of trying to rid himself of all received ideas and prejudices in order to reach the deduction of a few certain truths, he begins by asserting that he must hold up everything he has believed to be true to the most rigorous scrutiny. Much of what he believes to be true is dependent on the senses, and the senses are highly undependable. The problem is, however, where to stop in this process of doubting. The debate between Derrida and Foucault takes place over the following passage in Descartes, which follows the description of the doubt to which he will submit all sensation. Descartes puts a stop to the trajectory of doubting when he writes, "And how could I deny that these hands and this body are mine? If I were to do so, I would be like those madmen whose brains are so troubled and confused by black vapors and bile that they assert constantly that they are kings when actually they are very poor, that they are dressed in gold and silver when they are naked."[51] For Foucault, this is the moment of the inevitable exile and marginalization of madness. Derrida, however, reads on in the *Méditations* and finds that the next paragraph begins with *sed forte,* or what is translated in French as *toutefois,* or *nevertheless*—that is, nevertheless, Descartes will continue along the path of doubting and even arrive perhaps at denying that these hands, this body are his because he could be dreaming, or his sensations could all be the effects of a "Malin Génie," which is conspiring to deceive him in all things. Derrida finds that reason is always threatened by the sort of hyperbolic doubt that Descartes avails himself of in order to arrive at certain truths. The Malin Génie (or Evil Genius), after all, is the product of Cartesian reason and can only be overcome through a recourse to his double, God himself. The margins of reason are infinite, reason itself finite: God himself must intervene in order to secure the place of reason in relationship to its margins. Thus the margin, like the supplement, is that which is infinitely expandable. Madness, cast to the margins of the city of reason, marks the limits of reason by deferring the threat of its own ambiguous potential for proliferation. It is only in reading the footnote to reason, madness, that reason itself takes shape. Doubting is about living in two places, and therefore in between all places; Descartes secures a sense of permanent residence at the cost of submitting to a kind of house arrest. His sojourn in reason is threatened all the time by the possibility of a kind of internal exile. In representing the course of reason, Descartes has recourse to an experiment in doubting; Deckard as his uncanny cinematic double is trapped in this experiment.

Derrida reminds us that writing anticipates our eventual disappearance. Others die, but how can mortality be represented, except obliquely, by means of allegorizing the uncertainty about the moment when our eventual absence will begin? In the director's cut of *Blade Runner,* the film ends as Deckard and Rachel leave his apartment, intending to escape together toward an uncertain future.[52] The final words of the film, however, belong to Gaff, another member of the police force who favors dandyish Edwardian outfits and speaks in a guttural urban dialect. Gaff's line, "Too bad she won't live, but then again, who does?" refers to the built-in obsolescence of the Nexus 6 replicants, the cybernetic organism that destroys itself after a life span of four years. Gaff's question calls up the following response: no one lives forever. It also revives the earlier rhetorical question that haunts Deckard, "What's the difference?"

"What's the Difference?"

"Far from the machine being a pure absence of spontaneity, its resemblance to the psychical apparatus, its existence and its necessity bear witness to the finitude of the mnemic spontaneity which is thus supplemented. The machine—and consequently, representation—is death and finitude *within* the psyche."[53] The machine resembles the psychic apparatus while also figuring the limits of the apparatus. The replicants, as machines, are figures of and for representation. They also point toward the impossibility of intrapsychic representations of death itself. As such, they are figures for the impossibility of drawing the limits of the psychic apparatus, presenting us with the difficulties of representing *psyche* and the death of the psychic. The psychic is always already disappearing or fading into the nonpsychic: in the Freudian system, consciousness is always punctuated by moments of a radical absence of consciousness. These moments of occluded consciousness foreshadow nothing less than death itself, the absolute disappearance of psyche. "Too bad, she won't live, but then again, who does?" is finally an ironic statement on the impossibility of a happy or an *un*happy ending. If this question is understood as rhetorical (inviting therefore the negative response "no one"), we can understand Rachel's and Deckard's fate as neither happy nor unhappy. No one lives: the implied answer to Gaff's question haunts the closure of *Blade Runner*'s cinematic narrative. If the replicant succeeds as a copy of its organic original, it does so on the very ground of its own mortality.

It would seem that the resistance toward acknowledging the limits of literary "greatness" is a sign of our having thoroughly projected on it a dose of our own narcissistic delusions of immortality. The aura in which the literary text has been bathed, and on which it has subsisted, is the notion of an unchanging and immortal greatness. Critics like David Lehman try to defend the immortality of literary works from the destructive intent of literary theorists by making the crudest types of arguments for absolute difference: "Deconstructive critics are Nazis, we are not, we are lovers of literature." At the end of his book, Lehman exhorts us to support him in the following conclusions:

> Language by deconstructive decree is alien from human purposes, a stranger to human wishes and will. As a doctrine this is pernicious to the precise extent that it acquiesces in the curtailment of human freedom, for that is what is at stake in our ability to shape our words for our own purposes.[54]

The repetition of the notions of "human freedom and human purposes" does not make them any more meaningful: just as Bénichou's criticism of theory had recourse to an idealized notion of freedom and spontaneity, the affirmation of human will and human value is circumscribed by a series of radical exclusions. In Judovitz's critique of Cartesian method, the violence with which reason usurps the experience of the senses leaves the body to be one of philosophy's outcasts: this body returns to haunt reasonable discourses as the automaton and leaves its imprint on the commonly espoused denigration of both copy and machine. The machine, and the copy, is alien to human reason and must be cast out. Derrida, as well as de Man, tried to show that there were disturbing limits to human spontaneity and creativity with regard to language: their critique of the idealization of human reason and human will returns to the supplementary exclusion on which the human is always founded.

That this critique elicits powerful resistances is not surprising; there is however, a striking homogeneity in the language of theory's critics. The critique of theoretical criticism devolves to a series of clichés: it would seem that this antitheoretical discourse is hypnotized by the alien that it seeks to expel. The recourse to an idealization of the "human will" is automatic, highly predictable, mechanical, and repetitious. While these critics espouse freedom, they try to prevent thinking through the idealization of human will and human purposes. To make deconstruction the enemy of humanity seems to be the purpose of journalists and critics such as Lehman. This kind of discourse has left its mark on the terms of a public debate about the fate of the humanities. The irrational hatred of Paul de Man is just the other side of the irrational devotion that he inspired in his followers. Freud reminds us that transferential fantasies are difficult to undo: under the sign of late capitalism, the idealization of literature and its professors is, however, on the wane, helped along by both sides of the culture wars. But as the cultural prestige of literature has faded, so have grown more virulent the attempts to banish from literature's precincts any questioning of aestheticizing ideologies that attempt to deny the auratic extinction that never quite arrives but is always close at hand.

Chapter 3
The Princess of Clèves Makes a Faux Pas

IT IS INEVITABLE THAT IN THINKING THROUGH MACHINES, we cannot avoid focusing on the question of sexual difference. From Roentgen and Kintzing's dulcimer playing automaton to Hoffmann's Olympia and Villiers de l'Isle Adam's Eve, from Fritz Lang's Maria, to Phillip Dick and Ridley Scott's Rachel, the feminized or feminine technological double calls up a range of specific problems and urgent questions. In the next chapters, the issue of sexual difference becomes increasingly important because it becomes more and more clear that machines are feminized and identified with women in the following literary representations. I propose a return to an examination of the ways in which sexual difference has been dealt with in feminist literary criticism. The return to the text of feminist criticism from the past two decades is necessary because this work deserves to be read with greater care and attention. In revisiting the recent history of criticism, I continue to address the repression of the intellectual conflicts closest to us. It seems necessary, especially since this project addresses the many kinds of conflicts that have shaped the debates about the teaching and reading of literature.

Madame de Lafayette's novel *The Princess of Clèves* (1678) has inspired a particularly rich and heterogeneous body of criticism, but more important, certain feminist readings of this work in particular have inaugurated or even initiated a general feminist reevaluation of ancien régime texts. Because feminist criticism has acquired a kind of orthodoxy and discursive power, a serious and critical assessment of its intellectual strategies is now possible. The return to Lafayette takes place by way of a close reading of Nancy K. Miller's essay "Emphasis Added: Plots and Plausibilities in

Women's Fiction."[1] This essay will function as a "case" or an example (now much imitated) of a certain sort of feminist criticism. The Oedipal aspects of my engagement with Miller are obvious: she was my dissertation adviser, who was by turns, as most advisers are, inspiring and infuriating. Our intellectual differences were never fully articulated: an open disagreement with her would have led to an absolute break in the relationship, and so I avoided confronting the conflicts between our ways of working.

Intellectual inhibitions in feminist criticism have something to do with the fear of open conflict. Any kind of questioning of a feminist agenda appears to be a participation in the attacks on feminism. This is understandable in light of what the previous generations of women in academia had to struggle with in order to secure our positions in the university. This defensiveness, however, has led to an intellectual ossification and a particular orthodoxy in feminist studies. In my critique of Miller, I hope that I am not only acting out Oedipal conflicts: it should be obvious that my reading of her work is an attempt to pay tribute to and to honor her contributions. This is a challenge to feminist hermeneutics, but it is at the same time the recognition of the power of the feminist intervention. Feminist readings of *The Princess of Clèves* have attempted to work through the issues of the representation of sexual identity and desire in a context critical of an implicit refusal to deal with the question of sexual difference and narrativity in other forms of literary criticism.

In assessing the consequences of deconstruction's influence on the study of literature, David Lehman identifies feminist literary criticism as taking part in theoretical hegemony:

> The edicts of deconstruction—merged, to whatever extent, with the ideologies of Marxism, psychoanalytic theory, and feminism—remain the prevailing suppositions of the lit-crit establishment. One can discern a fundamental deconstructive procedure at work in the meteoric rise of "gender" and "ethnic" studies, at present the hottest areas in the lit-crit profession.... The profession's latest hotshots are still asking the question de Man lifted from Archie Bunker: "What's the difference?"[2]

Lehman might have put his finger on something here: literary critics may be divided into two camps. The first are the ones for whom the question has no meaning at all: they are contemptuous of it because for them everything is settled in advance. Literature is great; fads that are "hot" are bad; time will tell. For the second group, consideration of difference is productive and troublesome. How difference is constituted, defended, and mobilized is a question that can be posed nowhere else but in the humanities. The derision with which Lehman writes about what is "hot in the lit-crit establishment" is a sign of nothing more than resentment, and it is echoed in some of feminist criticism's most virulent critics.

We only have to consider Odile Hullot-Kentor's critique of Miller and Joan DeJean in her 1989 essay "*Clèves* Goes to Business School: A Review of DeJean and Miller."[3] While Hullot-Kentor enumerates a number of arguments that could be made against the assumptions behind feminist criticism (for instance, all women's writing is *essentially* subversive), she offers up nothing more intellectual than an attack on feminist ambitions. (Hullot-Kentor seems most irked by the fact that all that feminists seem to want is real estate.) The irony of such a sanctimonious position, especially in relationship to the seventeenth-century text, seems to be lost on Hullot-Kentor. She accuses Miller and DeJean of intellectual eclecticism and opportunism; she also accuses them of ignoring historical contexts. Hullot-Kentor's language, however, is marked by a peculiar relationship to her contemporary moment. She explicitly rejects the conditions of contemporary critical thinking, and yet there is a very breezy contemporary rhythm in her own writing. Thus we find arguments phrased in the following manner:

> Words become fads which for a brief period seemed to fashion the world to one's taste, but because they are no more than figures of taste, remain empty.... The result of this entrepreneurial philosophy of language is paradoxical: language can no longer resist the intentions to which it is subsumed but, at the same time, as if in revenge, it no longer holds any intention whatsoever and falls empty. (259)

Hullot-Kentor accuses Miller and DeJean of being fashion victims: in her view, the timelessness of great literature is compromised by the contemporary conditions of reading. She implies that she is in possession of a "full" language, resistant to "fads" and brimming over not only with authentic intentions but with meaning as well. She then goes on to accuse Miller and DeJean of reading *The Princess of Clèves* as not a novel but "an arbitrary system of signs." In this reading for the arbitrary, the feminist critics come up with new insights into the significance of sexual difference in the development of Lafayette's narrative. That such a reading might be really innovative, might challenge a traditional view of the novel, is not taken seriously for a minute by Hullot-Kentor. Like Bénichou and Lehman, she concludes that a theoretical reading compromises the full meaning of literary language, and yet their own readings come up with the same ideas over and over again: timeless greatness, human freedom, and so forth. The reduction here does not seem to take place so much in the work of the theorists as in the reading of theory by its critics, who manage to conjure up a parodic image of the theorist as a heavy-handed, ambition-driven zealot. What Hullot-Kentor, Bénichou, and Lehman offer as an alternative to theory's literary object is a discourse on human autonomy and literary greatness that they feel should never have been questioned in the first place. Any questioning of these categories is a submission to fads, fashion, frivolity, and will-to-power. It may seem tiresome to recapitulate the antitheoretical position, but it is in its contemptuous repetitiveness that we can see clearly how intolerantly a view of literature and literary studies is being espoused.

The case of feminist literary theory is both more and less complicated than Hullot-Kentor imagines, and her neglect of critical reading leads her to impasses where we shall decline to follow. Suffice it to say that Miller and DeJean deserve a better reading and a better critique. Miller's essay will be our primary focus because it has been so influential, not only in the study of Lafayette's novel but also in the establishment of certain parameters of analysis for the study of "women's fiction" in general. Miller's reading of Freud is particularly important here: her stance is exemplary of a certain kind of politically driven impulse to correct and overcome psychoanalytic theory.

Will-to-Power and the Problem of Representation

Friedrich Nietzsche is the missing link in our return to a reading of *The Princess of Clèves*. By following the itinerary of *The Will to Power* (1901), we can arrive at an understanding of how Lafayette's *Princess* functions within a Nietzschean genealogy.[4] For Nietzsche, seventeenth-century France is the site of the blossoming of will-to-power; at this time the deployment of unsentimental force reached a high point of refinement before being corrupted by bourgeois constraints. Despite his critique of La Rochefoucauld as being caught up in Christian ideals, he finds in the uncompromising lucidity of the seventeenth-century moralist, a kindred spirit.[5] The question of will-to-power and its representations forms and informs the relationships between all the characters of Lafayette's fiction, especially in *The Princess of Clèves*. Moreover, the representations of exemplarity and inimitability in this work allow for a theorization of the irreducibility and instability of differences between the world of appearances and the order of truth. What follows is a reading of the Princess of Clèves's renunciation as a moment of will-to-power: she produces and is produced as a successful example of perfect inimitability. In this way, the most precious of classical ambitions is fulfilled in the topos of Lafayette's novel.

How the princess arrives at her will-to-power, however, is a narrative problem that is complicated by the conventions of representation itself *(vraisemblance, bienséance)*. In a Nietzschean context, the seventeenth-century subject (as it has been represented in this fiction) appears bathed in a chiaroscuro of Baroque metamorphoses that invites the radical negation of representability: interiors and exteriors transform themselves into one another without necessarily communicating with one another. The topos of Baroque subjectivity can be considered from a contemporary point of view as an antipode to ideologies of transparence and presence that Nietzsche criticizes so vehemently in the form of Rousseau's sentimentalization of virtue. Feminist reappraisals of Lafayette's fiction have focused on the ways in which it represents what might be called a heroics of femininity; what I would like to do here is to work through how a heroism that only affirms will-to-power must always disguise itself behind many different veils and produces a particularly demanding set of representational and hermeneutic problems.

Thinking through Lafayette's novel in the context of will-to-power allows us to reevaluate the heroine's enigmatic renunciation of passion. If the princess is marked by traits that distinguish her from all others, the novel that represents her is one that aspires to the same originality. The princess hides her will-to-power behind a veil of renounced passion (erotic wishes). The author veils her ambition (egoistic wishes) behind anonymity. If the princess renounces her passion, this spectacular renunciation produces the figure of the inimitability of her virtue. Her will-to-power (being inimitable) is perfectly aligned with her renunciation, and the representation of such a wonderful coincidence is the implicit ambition of the author. The princess's retreat can be read as an inverted figure of Lafayette's own anonymity as an author. The princess's virtues, described in the beginning of the novel as hyperbolically inimitable, become in the end, conventionally unrepresentable.

Writing Wrongs

Feminist criticism of *The Princess of Clèves* has mobilized itself around the problem of righting the wrongs that have surrounded its reception: this novel and its heroine have been "misunderstood." Miller's "Emphasis Added" deals with Bussy-Rabutin's critique of the novel's *invraisemblance* (its relationship to *bienséance*), especially the question of the princess's confession to her husband of her adulterous passion. Miller shows how Bussy-Rabutin's critique of the novel is ideologically motivated; in doing so, she protects and rescues the novel from Bussy-Rabutin's criticisms. In this context, Miller corrects Freud's 1908 essay "Der Dichter und das Phantasieren." Her corrections begin with an emphasis on the deployment of sexual difference in the Freudian text; instead of referring to the *Standard Edition* of Freud's essay, she cites a 1925 translation of the text titled "The Relation of the Poet to Day-Dreaming" (in the *Standard Edition* translation, the title of the essay is "Creative Writers and Day-Dreaming"):

> The impelling wishes vary according to the sex, character and circumstances of the creator; they may easily be divided, however, into two principal groups. Either they are ambitious wishes, serving to exalt the person creating them, or they are erotic. In young women erotic wishes dominate the phantasies *almost exclusively,* for their ambition is *generally comprised* in their erotic longings; in young men egoistic and ambitious wishes assert themselves plainly enough alongside their erotic desires.[6]

I would like to cite the *Standard Edition* translation of this text and add a sentence that Miller's citation significantly omits. It would seem that its inclusion would have led to greater complications than she is willing to take on in her argument against Freud. Freud refers to the motivation behind fantasies as being one that is based on imagining the fulfillment of a wish and the correction of an "unsatisfying reality":

> These motivating wishes vary according to the sex, character and circumstances of the person who is having the phantasy; but they fall naturally into two main groups. They are either ambitious wishes, which serve to elevate the subject's personality; or they are erotic ones. In young women the erotic wishes predominate almost exclusively, for their ambition is as a rule absorbed by erotic trends. In young men egoistic and ambitious wishes come to the fore clearly enough alongside of erotic ones. *But we will not lay stress on the opposition between the two trends; we would rather emphasize the fact they they are often united.*[7]

By "laying stress" exactly where Freud urges his readers not to, Miller is able to point to a wrong that needs to be righted—Freud's attribution of ambition to the young man, and erotic wishes to the young woman. Indeed, she does go on to perform an act of feminist revision and correction. In placing an emphasis on the absolute and unmovable nature of sexual difference in Freud, she makes her task as a feminist revisionist simpler.

We should also be aware of the specificity of the German term *Ehrgeiz*, translated in the *Standard Edition* as ambition: it can also be understood quite literally as a kind of greediness for honor, an avarice for respect and awe. *Ehrgeiz* is almost an oxymoronic conflation of two words, *greed* and *honor*, that cancels out the higher value of the latter. *Liebesstreben* is translated simply as erotic wishes, when erotic aspirations would also be an accurate translation of the German term: *streben* implies a sense of striving, an activity slightly different from the passivity of simply "wishing." In these erotic daydreams, then, Freud finds in a young woman's striving for love a reaction to the constraints on feminine sexuality: "The well-brought up young woman is only allowed a minimum of desire"; her will for autonomy and freedom, her ambition, therefore, might very well be colored, first and foremost, by a striving to overcome the minimal portion of desire that has been parceled out to her.[8]

For Freud, ambition is always already eroticized: for the young man in the Freudian model, ambitious fantasies are often based on his need to impress a woman. Freud sets up a contrast between two kinds of daydreams, one erotic and the other ambitious, differing insofar as they are marked by the gender of the dreamer, but he is forced to conclude that there are erotic impulses in the ambitious dream of the young man as well:

> Just as in many altar-pieces, the portrait of the donor is to be seen in a corner of the picture so, in the majority of ambitious phantasies, we can discover in some corner or other the lady for whom the creator of the phantasy performs all his heroic deeds and at whose feet all his triumphs are laid.[9]

The lady can be discerned only with great care because she is hidden, integrated into the picture (the daydream), as one might integrate the image of a patron or donor who has paid for an altar-piece and whose gaze the final painting seeks to flatter, albeit discreetly. There is no ambition that does not contain an erotic impulse. Miller reads the eroticization of ambition in the young man's daydream as another case of Freud's giving the young man both ambitious and erotic impulses, while leaving the young woman bereft of ambition and only in possession of striving for love. Miller goes on to question whether or not there is an ambitious impulse hidden in the erotic wish of the young woman, but this reversal of the opposition is based on a need to establish an impossible symmetry between the sexes.

Freudian "reality" is unsatisfying and always already in need of revision for both men and women. Erotic wishes are always thwarted by something in reality. Thwarted erotic impulses can express themselves as ambitious daydreams; in fact, the pleasure of the ambitious impulse depends on its libidinal origins. For psychoanalytic theory, erotic impulses are primary impulses. Ambition for Freud is only a secondary phenomenon. It is the transformation of an infantile, erotic formation, usually associated with an infantile overinvestment in urethral pleasure.[10] An ambitious impulse is always secondary, and in this sense always a deformation of an erotic impulse. Erotic aspirations are transformed into ambitious ones as polymorphous infantile sexuality is given structure, but never fully mastered. In this analysis, *Liebesstreben* is always aligned with *Ehrgeiz:* the greed for honor seems less honorable, while the striving for love seems more ambitious.

Miller accuses Freud of glossing over the significance of sexual difference in two instances. The first has to do with the fact that Freud "conjures up only a male creator" (40); Miller criticizes Freud for only being able to imagine a male writer. Second, in Freud's analysis of the formulation "Nothing can happen to me!" as the pivot of identification between reader and literary hero, there is also no account of a heroine in relationship to a female reader. For Freud, "Nothing can happen to me!" is the ego's motto: the ego recognizes itself in the male hero. Miller reminds us that there are literary heroines in French women's fiction who also triumph by rising above it all. All this appears to be a pertinent critique of Freud's neglect of feminine ambition. Miller writes that we find in women's fiction cases where

> the heroine proves to be better than her victimizers; and perhaps this ultimate superiority, which is to be read in the choice to go beyond love, beyond "erotic longings," is the figure that the "ambitious wishes" of women writers (dreamers) takes. (41)

The heroic overcoming of erotic longings leads us to conclude that for Miller there is something implicitly negative about the erotic impulse. The ambition "to go beyond love" is an ambition that gives women writers pleasure. In a strictly psychoanalytic sense, however, the relationship between erotic desires and ambition is not an oppositional one: sexual desire actually gives rise to the drives of ambition along primordial and onanistic paths. For psychoanalytic theory, ambitious impulses are always already erotic. Miller suggests that there can be a fantasy of power that is corrective and revisionary (we shall restore to women a fantasy that they have been wrongfully denied, we shall "revise" the patriarchal grammar in which texts have been written). This particular form of ambition is completely purged of all erotic content. According to Miller, this is the ambition that is to be found in the work of "women writers" where

> egoistic desires would assert themselves paratactically alongside erotic ones. The repressed content, I think would be, not erotic impulses, but an impulse to power: a fantasy of power that would revise the social grammar in which women are never defined as subjects; a fantasy of power than disdains sexual exchange in which women can participate only as objects of circulation. The day-dreams or fictions of women writers would then, like those of men, say, "Nothing can happen to me!" But the modalities of that invulnerability would be marked in an essentially different way. I am talking, of course, about the power of the weak. . . . (Perhaps we shall not have a poetics of women's literature until we have more weak readers.) (41–42)[11]

Miller has led us to the identification of absolute difference between the writing of the two sexes, marked "in an essentially different way." When Miller refuses the instability of sexual difference and embraces instead the essential differences between the male-authored and the female-authored text, she attempts to mitigate the radicalness of psychoanalytic theory by positioning women as authors of the more significant epistemological break with forms of knowledge and constructions of gender. It was Freud, after all, who taught us that such an impossibility was based on the constitutional bisexuality of the subject,[12] but this theory has provoked a great deal of discontentment.[13] Sarah Kofman criticizes Freud on this point, but she does not fail to bring out the significance of the epistemological break that psychoanalysis makes with philosophy and popular opinion in its treatment of sexual difference. Freud's position on the formation of sexual difference in the essay on "Femininity" (1932) is adamantly opposed to what Kofman calls "the pseudocertainties of popular opinion" about the absoluteness of sexual difference. Freud's thesis of bisexuality may favor masculinity—it may even reformulate a hierarchy of difference between masculine and feminine traits—but it refuses to give in to certainty. Kofman writes that Freud appeals to science in order to disrupt a purely anatomical and commonsensical notion of sexual difference:

His recourse to science does not have as its goal the quest for security; on the contrary it aims at plunging into aporia: "You are bound to have doubts as to the decisive significance of those elements and must conclude that what constitutes masculinity or femininity is an unknown characteristic which anatomy cannot lay hold of."[14]

A feminist critique of Freud that does not take this into account will find itself defending a regressive position, and one in which sexual politics is constructed out of essential differences between men and women. Put crudely, Miller's position sets up the sexual anatomy of the author as the determining instance of a literary hermeneutic.

Freud opens "Creative Writers and Day-Dreaming" with a deceptively naive question, "We laymen have always been intensely curious to know ... from what sources that strange being, the creative writer, draws his material" (*SE* 9:143), as well as the answer that he manages to formulate in response to it. Children play, adults daydream (in private), while creative writers do it in public. For Freud, the creative writer is "a dreamer in broad daylight."[15] Writing novels is a kind of public daydreaming, then, but Lafayette as an author eclipses herself so that she can enjoy what Joan DeJean calls the privileges of anonymity.[16] The novel describes the princess daydreaming in front of a portrait of Nemours. While the content of her reverie appears to be entirely erotic, the form of representation that contains it (the novel) is entirely ambitious.

"The motive forces of phantasies are unsatisfied wishes, and every single phantasy is the fulfillment of a wish, a correction of unsatisfying reality" (*SE* 9:146). Fantasizing, then, is always already a kind of rewriting or a correction of reality, a correction and a revision of an endlessly unsatisfying text or manuscript. In her reverie, the princess is "correcting" her reality, and in her fiction making, Lafayette is revising hers. Following the passage on the differences between the wishes of young women and men, Freud hastens to warn his readers against assuming that fantasies that make up daydreams "are stereotyped or unalterable." In fact, they are constantly changing according to the differing conditions of the subject's "shifting conditions of life." The Freudian fantasy changes and shifts, hovering above linear temporality (between past, present, and future) so that fantasies are always already revising themselves even as they seek to "correct" reality.

However, the pleasures of reading are described by Freud as having something to do with the creative writer's ability to soften "the character of his egoistic daydreams by altering and disguising it, and he [the writer] bribes us by the purely formal—that is aesthetic—yield of pleasure which he offers us in the presentation of his fantasies" (*SE* 9:153). By formalizing and aestheticizing daydreams, the writer gives us a taste of "fore-pleasure."[17] The enjoyment that an imaginative work procures for us has something to do with the "liberation of tensions in our minds." Freud concludes that reading "creative" writing produces pleasure because it allows us to enjoy our daydreams with less shame. Pleasure in reading can be achieved only because the very structure of the reader's daydreaming (in private) has been displaced and transformed so that the reader is participating in the writer's egoistic display (in public).

The pleasure of Lafayette's daydreaming is formalized in the renunciation of her princess. Thus reading and writing are activities that mix private pleasures with public display: they permit us to daydream between private and public spheres, and in so doing, as specific activities, they provide a very specific pleasure. In a feminist context, when the "liberation" of a woman's ambitious impulses takes place in writing, she can finally enjoy a precious fantasy: that the realization of egoistic impulses can coincide with the achievement of a political progress. Feminist criticism that is engaged in the liberation of female authenticity founds itself on a fundamental and radical differentiation between "male-authored" texts and "female-authored" ones. How does this notion of authenticity fit in with other ideologies of authorial presence and sexual difference? If the author has been deposed as the last instance of authority on the interpretation of the text, this leaves her as a haunting presence, which we cannot do away with altogether either. In both cases of de Man and Lafayette, a certain amount of biographical information offers us important insights into reading their work. Is there a kind of spectral authorship that haunts every text and every reading, and can a reading for such hauntings allow us to elude the double traps of biographical criticism?

Hullot-Kentor does identify something disturbing about a feminist reading of canonical texts: its newness produces an appearance of arbitrariness. It seems that one of the conditions of a theoretical orientation toward literature concerns an acute awareness of the arbitrary or nonessential nature of convention, especially conventions of reading the literary work. One of the most important of these conventions is based on reading a work as simply determined by its author's biographical experiences. The occlusion of the author has led, however, to a singular uneasiness about how to stabilize textual meaning: by evoking the author's sexual identity as a condition of reading, a new form of authorial presence is evoked.

If there reigns a singular confusion in literary studies (specifically in analyses of ancien régime texts), it could have to do with a rupture in the literary *epistēmē* around the relationship between truth and appearances. The grounding of a literary representation of "true being" gives way, especially under theoretical pressure: merely condemning the work of the thinkers who have produced this rupture is not going to make things return to the way we thought they were. In Derrida's reading of Rousseau, authenticity becomes one of the problems around which the difference between writing and speaking turns. Derrida equates "art, *technē*, image, representation": all of these are different forms of supplementarity and figures for writing with convention. Derrida's reading of Rousseau gives us a way of identifying in the writing and the thinking of the latter a formation of vital importance: convention, representation, and *technē* all take part in the work of supplementarity. When these terms attach themselves to presence, they diminish its authenticity. *Bienséance* masks the vices, but vice, for Rousseau, is nothing more than the mask itself.[18] Rousseau condemns the mask: the mask is always doing the work of deception. When Nietzsche affirms cunning, ruse, pride, and artifice, he breaks with a genealogy of bourgeois authenticity and returns to a seventeenth-century position, embodied by La Rochefoucauld. For Nietzsche, "everything deep loves masks."

To understand how feminine virtue is not all that it seems, especially in a literary setting where plays for power are being made, we turn to Rousseau's *Julie: The New Héloïse* as an alternative narrative of feminine renunciation. In it, Wolmar says to Saint-Preux when the latter addresses Julie as Madame: "*Bienséance* [convention] is nothing more than a mask for the vices: where virtue reigns, *bienséance* is unnecessary."[19] When Saint-Preux and Julie are reunited under Wolmar's paternalistic gaze, Saint-Preux is careful to address Julie as "Madame," but for Wolmar, this title is a mark of the unnecessary constraints of formality. Where there is virtue, forms are rendered useless. "Madame" puts a respectful distance of form between Julie and Saint-Preux; the title acknowledges Julie's married state, Julie's privileged relationship with another man. However, as Saint-Preux testifies, the virtue that reigns and will continue to reign in this situation is guaranteed by Wolmar's *authority* rather than Saint-Preux's formality. It is Wolmar's authority that guarantees Saint-Preux's new virtue. Under this new regime, where Wolmar is transformed into a presence that is both ubiquitous and invisible, there is no need for the distancing effect of forms and formalities.[20] Feminine virtue stands as the guarantee of paternalistic authority, a force so powerful that all forms and conventions are abolished.

Convention for Rousseau is false: it is nothing more than the mask itself. As a mask, it marks the difference between being and representation. In Rousseau, *bienséance* is always under suspicion of staging a cover-up. It is falseness trying to pass itself off as truth. Truth here is what is dissimulated, what is not to be revealed, vice itself, in this case the adulterous desire between Julie and Saint-Preux. For Rousseau, forms of social dissimulation actually produce vice. To dispose of formalities is to destroy the vice that they seek to conceal. The truth of *bienséance* is that it masks vices, puts on a good face in order to pass off viciousness for virtue. In Jean Starobinski's terms, *bienséance* would be, then, one of the most important obstacles to transparence. His reading of Rousseau's *Discourse on the Sciences and the Arts* emphasizes the fact that Rousseau's obsession with the corrosive force of the differences between appearances and truths, while not original for its time, laid out a lifelong preoccupation with this problem: "That appearance and reality are two different things and that a 'veil' covers our true feelings—this is the initial scandal that Rousseau encounters, this is the unacceptable datum for which he will seek the explanation and cause, this is the misfortune from which he longs for deliverance."[21] Appearance itself is such an obstacle because appearances are formal; appearances that have been arranged according to social convention *(bienséance)* can be nothing more than pure deception. Virtue for Rousseau is a virtue that develops from *within* and is not the product of either external coercion or threats of physical violence. It destroys representation. Innate virtue, the virtue of Émile, for example, develops organically rather than synthetically and is capable of breaking down the very difference between inside and outside: as such, it is capable of overcoming space and distance itself.

In past readings of Lafayette's novel, the ideal of transparence gives structure and form to a system of values on which the interpretations are based, and the Princess of Clèves is read as the harbinger of a new standard of virtue. In *Le Roman jusqu'à la revolution*, Henri Coulet writes that "*bienséance* does not forbid Madame de Clèves to love Nemours; it simply forbids her to show it! . . . Madame de Clèves rebels against *bienséance*, a system of dissimulation and lies; she is committed to transparence and not to representation; it is to sincerity that she sacrifices herself."[22] Coulet privileges transparence and sincerity in relationship to *bienséance* (or representation), which are equated with dissimulation. Madame de Clèves's commitment to sincerity is not in fact so clear: if she were devoted to nothing more than sincerity, why does she refuse to marry Nemours after her husband's death when she is absolutely free to do so? If *bienséance* is nothing but a system of false appearances and dissimulations, what is the sincerity to which she sacrifices herself here? Would not a marriage of love be the ultimate act of transparency and sincerity? The princess does not so much opt for transparency over representation but, rather, finds recourse in an enigmatic asceticism. It is a code that would not permit her to marry a man she loves and that, like the forms of *bienséance,* demands a disjunction between what she feels and how she acts. This disjunction is based on the possibility of the intervention of both artifice and cunning: it points toward the mask, which covers over the break between experience and representation. When the novel opens, the princess is thrust on stage: both theatrical and theoretical, she defies the very sincerity with which Coulet would like to endow her, and in taking on her role with great seriousness, she plays one of the greatest acts in what becomes one of the first modern novels.

Joan DeJean describes the classical novel as the construction of surfaces, but more than that as a machine that controls and regulates all readings:

> The dazzling surfaces of the artistic creations that serve as models for the Sun King's age constitute a brilliant machine for controlling all readers, for keeping them in the dark, for discouraging them from asking questions about the identity of the master artist who surrounds them with "dorures" [gilt] to protect his monstrous invisibility.[23]

DeJean's description is charged with the tensions of a series of implied oppositions: surface/depth, inside/outside, control/freedom, decor/functionality, visible/invisible, presence/absence. According to her, the structure of the classical novel is an allegory of Louis XIV's form of authority: as a machine of surfaces, designed to hide the authorial presence, the presence of the master/author. The novel's secret is as well guarded as a Vaubanian fortress itself. For DeJean, as for Rousseau, the surface is also necessarily and constitutively deceptive. The truth lies in the hidden depths of the fortress and at the heart of the machine. The truth involves a "monstrous invisibility" and a corrosive emptiness.[24] For DeJean, as for Rousseau, surfaces, like appearances, are always deceptive, always dissimulating a lack and disguising an emptiness. The machine recurs here as a figure of deception and complexity, as an instrument of the tyrant's control. What is being manipulated by this machine is specularity itself: the surfaces are so many glittering mirrors whose reflections dazzle and blind the viewer, the reader, and the subject.

When Rousseau asserts that true happiness can only be achieved when the external countenance is an image of the internal disposition of the heart, he wants to secure a privileged relationship of specularity: "How sweet life would be for us if the external countenance was always an image of the heart's dispositions."[25] The face should always be a mirror of the heart. That the face is not produces a radical noncorrespondence between inside and outside: this disjunction is at the root of all unhappiness and evil in Rousseau's world. Representation, which is on the side of the outside, on the side of the surface, leads to vice itself. The interior is in fact impoverished by the exterior. A void forms inside. The novel establishes a space in which it is possible to represent the noncommunication of two interiorities: the interior life of the princess is doubled, unknowable to herself, as it is hidden from others. The nonspecular relationship between inside and outside creates the possibility of writing: writing occurs in the disjunction between being *(être)* and appearances *(paraître)*. Dissimulation sets off the possibility of a proliferation of an endless regression: there are dissimulations within dissimulations.

For Rousseau, masks are the index of superficiality: for Nietzsche, they are the signs of a recuperation of a temporality that destroys any grand narrative of history and historicity: the passage of time does not reveal deeper truths. In this conception of generalized inauthenticity, everyone is a player in the field of theory and theater. This is the conception of history that Walter Benjamin offers us in the *Origin of the German Tragic Drama*.[26] The tragic dimension of the Baroque period is evoked by Benjamin in his reading of seventeenth-century German plays whose representations of life at court offer a glimpse into the way in which submission and powerlessness are staged as cunningly wrought tableaux. Benjamin tries to read a literature caught in the field of the tyrant's court, and the provocative associations that he makes between tyranny and feminine asceticism offer us a way out of the impasse of the classical oppositions between appearances and truths:

> The function of the tyrant is the restoration of order in the state of emergency: a dictatorship whose utopian goal will always be to replace the unpredictability of historical accident with the iron constitution of the laws of nature. But the stoic technique also aims to establish a corresponding fortification against a state of emergency in the soul, the rule of emotions. It too seeks to set up a new, anti-historical creation—in the woman the assertion of chastity—which is no less far removed from the innocent state of primal creation than the dictatorial constitution of the tyrant. The hallmark of domestic devotion is replaced by physical asceticism. Thus it is that in the martyr-drama the chaste princess takes pride of place.[27]

The Princess of Clèves is in many ways the most chaste of all princesses, the one who best manages the state of emergency in her heart by cutting off all possibility of peaceful succession in her relationship to men, passion, and marriage. The immolation of her desires is an allegory of the heroic mortification of ambitions, both erotic and political. The restoration of peace and harmony in the tyrant's court comes at a very high price: the fragile political stability that we can compare with what Domna Stanton has analyzed as the seventeenth-century ideal of "repos" is constantly under threat.[28] It is no wonder, then, that a woman's powers of ascetic renunciation would have fascinated those sharp readers, those powerful men and women of the seventeenth century, who turned about the Sun King like so many careful satellites and were witness to one of the most radical consolidations of power that Europe had yet known.

The Princess Is a Palimpsest

The Princess of Clèves is a novel that is constructed according to a logic of superlatives. It opens with a famous one: "France had never seen such Magnificence and gallantry as in the last years of the Henri II's reign."[29] Every portrait the narrator paints is a flattering one, but two characters are described in terms of the unrivaled—the Duc de Nemours and Mademoiselle de Chartres. Their love is figuratively overdetermined from these opening pages: they are exemplary members of an exemplary court whose qualities, like theirs, are distinguished by the superlative. Nemours is described as possessing such an extraordinary combination of virtues, qualities, and beauty that he embodies something that has never been seen before (243). Mademoiselle de Chartres is distinguished by a quality of absolute unprecedentedness as well: "The paleness of her coloring and her bold hair gave her an air that one had never seen before her."[30] Lafayette's superlatives construct a paradigm of incomparability that is applied to Nemours and the princess.

We could say that all of Lafayette's heroines are involved in a kind of endless and impossible regime of disciplining the self, its transgressive desires, its involuntary movements, and the threatening chaos of its passions.[31] The Princess of Clèves's radical singularity is established from the beginning of the narrative, but it is consummated in the singular confession. Joan DeJean offers this reading of this confession in "Lafayette's Ellipses: The Privileges of Anonymity" as a crucial moment in the princess's coming-to-authorship or self-representation. It is she who, in reflecting on what has transpired, defines her own act: "The singularity of such a confession, for which she found no examples, made her see all the danger that threatened her."[32] For DeJean,

> the Princess' *aveu* [confession] is the text of her self-definition, the mark of her self-constitution, her signature. The Princess exchanges her husband's estate for the female literary estate. . . . The Princess, like her creator, replaced the acknowledgment in the male script *(le nom d'auteur)* with a signature in the feminine.[33]

The promotion of a feminine signature as self-defining and self-constituting is convincing if we do not consider the complexities of renunciation under the conditions of absolutism.[34] For Benjamin, the bold asceticism of a chaste princess is only one of an entire constellation of qualities that she must embody in order to have pride of place at the tyrant's court. The feminine signature is one that seals for DeJean, as for Miller, the feminist parable in this text. But renunciation, however heroic, does not guarantee that a "female literary estate" is a space free of domination or repression.

The subject makes herself recognizable through submission to the conventions of *mondanité,* or worldliness: this kind of self-mastery is acquired through the discipline of constant dissimulation and vigilance.[35] The signature is testimony to the subject's submission to convention. It can also be read as being the mark of a writer's constant negotiations with constraint. At court, every good courtier seeks ways of distinguishing himself through the creation of a perfectly refined "conventional mask." The submission to *bienséance* sheds light on the dilemma of the seventeenth-century author in the following manner: how does one write according to the code while distinguishing oneself as singular and "original"? These are the questions of a subjectivity allegorized as authorship: how does one submit to the conventions of genre, usage, and grammar while distinguishing oneself in one's writing as unique and original? How to acquire distinction through discipline? To write is to "advance" as masked *(larvatus pro deo),* signing on the dotted line of one's own duplicity by entering into a contract with the conventions that guarantee a certain legibility.

DeJean implies in a convincing way that Lafayette's own preciously guarded anonymity comes to be allegorized in the novel; the princess, who is author of nothing but herself, succeeds in encrypting her desire in the famous *pavillon* while the author succeeds in encrypting her identity in the narrative of the novel. Cunning acts of dissimulation through renunciation, they have been read by feminist critics as heroic acts of escaping from "male control." Reading for "signatures in the feminine" becomes a critical act that for feminist critics implies the marking of domination and control as irreducibly masculine.

Civilization and Its Discontents

Coulet describes *bienséance* as an imperative to suppress all subjective particularity and sentiment: *bienséance* can be analyzed as a set of cultural injunctions that have a superegoic function. The jurisdiction of *bienséance* does not extend to the inner life of its subjects: it prescribed that all signs of individual particularity be transformed into conventional forms and behaviors, but it pretended to do no more than that. The imperative that Freud focuses on in *Civilization and Its Discontents* (1939) is the Christian commandment to "Love thy neighbor as thyself."[36] He compares the new demand to the First Testament imperative "Thou shalt not kill," which only seeks to govern external gestures. What the Christian injunction aspires to govern is nothing less than the inner life of the subject. The commandment seeks to banish aggression and its detrimental effects on the human community. The impossibility of complying with this commandment, according to Freud, only makes it seem more valuable. It is here that Freud and Nietzsche are in uncanny agreement, for Nietzsche reminds us that because the moral injunctions of Christianity are impossible to comply with, they are in fact designed not to make human beings be more virtuous but to make them feel more sinful.

There is a relationship between the psychoanalytic notion of a constitutive aggressivity and Nietzsche's will-to-power that can be understood as being deployed across the double spheres of action in Lafayette's novel—*l'ambition et la galanterie*: "Ambition and gallantry animated life at this court, and they occupied men as well as women in equal measure. There were so many different cabals and interests, and the women were greatly involved because love was always mixed with intrigue and intrigue with love."[37] It is impossible in this court to separate *ambition* from *galanterie*. It is the "here," the "at this place," or "in this court" that designates a space and a place in which love and ambition are inseparable.

Lafayette's heroine is torn between the aggressive desire to distinguish herself (*ambition* and *galanterie*) and the civilizing imperative *(bienséance)* to suppress all marks of particularity or sentiment. The tensions that traverse the subject of such a civilizing process are represented in the narrative of the novel: the novel becomes the site of their representability. In

Lafayette's text, we find that the very sublimation of individual particularities into social forms produces another kind of aggression, this time directed inward in the form of self-surveillance and outward in the form of writing. The novel becomes the form of representation for this complex form of internalization: this literary form represents the internalization of conscience that produces virtue: in other words, the most powerful and effective injunctions are intrapsychic rather than intersubjective.[38] There is nothing outside of the princess herself that prevents her from accepting Nemours as a lover or a husband at the end of the narrative. All external obstacles have been dissolved.

Norbert Elias uses the confessional scene between husband and wife in this novel as an example of the power of internalized discipline. After confessing to her husband her adulterous love, the princess pleads with him to shut her away somewhere, to keep her away from all society. The prince replies:

> No, Madame, . . . I want to trust only you; this is the path that my heart as well as my reason counsel me to take. With your character, in leaving you with your liberty, I give you narrower constraints than I could ever prescribe for you myself.[39]

No judgment or punishment will be as strict as that which can be expected from within. The prince leaves his wife alone with her conscience because he understands her particular character, consisting of an inimitable self-discipline. "Je ne me veux fier qu'à vous-même" can be understood as "I want to trust only you" as well as "I want to entrust myself only to you." In either case, the prince is unable to live by this statement and resorts to sending someone to spy on his wife. Elias fails to comment on the fact that the ambiguous report filed by his spy proves fatal: transparency in marital relations was not enough. The prince is incapable of entrusting his honor to his wife, without recourse to a bit of duplicity that ends up breaking them up for good. If the Prince refuses to give his wife a clear order, it is perhaps because the education she received from her mother consisted of nothing more than prohibitions: a long series of "il ne faut pas" or "one must not," followed by infinitive verbs all having to do with falling and loving. The lessons of the mother are internalized by the daughter and take on a life of their own. What the princess learns is that the tranquillity of a virtuous life is more desirable than the dangerous excitement of a passionate one.[40] Madame de Chartres is the agency of the civilizing process, but an external agency of punishment and authority is not sufficient to guarantee that the individual adheres to the terms of renunciation; this is why she must die in order for her injunctions to be fully incorporated in her daughter's discourse. The body of the one who forbids expires while the order to renounce is folded into the body of the daughter herself.

At the end of the novel, when Nemours appears before the princess to ask for her hand, she replies to him in the spirit of her mother's lessons. Men are little capable of sincerity, they are unfaithful, "but can men preserve passion in this eternal bond? Can I hope for a miracle in my favor and should I put myself in a position to witness the certain end of a passion on which I would have based all my happiness?"[41] Madame de Chartres's lessons have been successfully integrated by her daughter into her own thinking. There is nothing that Nemours can say or do that will dislodge the mother's message from the daughter's reply. The injunction comes from the mother on her deathbed. Madame de Chartres makes a demand in no uncertain terms for her to never give into her passion for Nemours: "You are on the edge of a precipice: a great effort and a great violence will be necessary if you want to hold yourself back."[42] In other words, "Faut pas faire de faux pas"—it takes only one false step to precipitate the fall into the abyss of passion. In trying to avoid the fall, the princess manages to recover from more than just a brief stumble. Her refusal of Nemours's proposal is a sign that in successfully incorporating the mother's (and the husband's) injunctions, she is never going to get over having been guilty of an adulterous feeling at the same time that she is never really going to get over the double deaths of mother and husband. She commemorates their deaths by enacting her melancholic internalization of their prescriptive voices.[43] Thus she refuses to move on or let go, and everything will continue according to *bienséance*. Internationalization of proscriptive orders is always a sign of what Freud would call excessive mourning: this kind of obedience then anticipates the death of the voice of interdiction and at the same time exceeds the moral demands of that figure. In this way, the subject of Christianity obeys the impossible demand by going too far in an act of melancholic refusal. That this is dramatized by a novel should be no surprise: the novel is the medium and the mediator of the modern subject's negotiations with building (or *Bildung*) an inner life.

Earlier in Lafayette's novel, a premonition of how the princess is going to stage her withdrawal appears. After being approached by Nemours during a brief, unobserved moment between the comings and goings of the Reine Dauphine, the king, and his mistress, Madame de Clèves tries to avoid him ("elle le fuyait"): "Because the room was crowded, she tripped in her skirt and made a faux pas: she used this pretext to leave a place where she did not have the strength to stay, and pretending to be unable to stand, she returned to her rooms."[44] The faux pas becomes a pretext for the need to retreat to stillness. Stillness is alluded to by the *séant* in *bienséant*, which translated literally could mean "that which sits well." Caught by the pull of the precipitation and precipices (falling into her dress and falling into love), an *abîme* into which her desire, like gravity, threatens to pull her, the princess tries to remain seated. Such a step into or toward flight ("elle le fuyait") leads her to a faux pas: it brings her closer to Nemours while tearing her away from him at the same time. It is a *rapprochement* that leads to a distancing or *éloignement* at the same time: an ambiguous step certainly, one that affirms and negates at the same time. It is a step that is a stumble, a step that turns into a fall, which in turn makes it false— a false step, however, that reveals the truth. Derrida writes in the context of a reading of Blanchot's fiction, "Pas est la Chose."[45] The *pas* leads to the impossible literary thing: the faux pas is the one step that leads toward a fall into love that is always suspended between terrifying proximities and painfully long distances.

The strange *pas* can be read as the *pas* of *passion*. Passion intervenes as a kind of affective cataclysm that threatens the subject's very mastery of social functions and symbolic positions. This is the state of emergency to which the tyrannical imposition of interdiction is opposed. Passion disturbs the smooth surfaces of the decorous mask and manifests itself in the form of slips, mistakes, and lapses of attention.[46] It invites the *mis*take or the *mis*step that transforms itself into the pleasure of the fall, the giving into the gravitational pull of precipitation and precipices. The faux pas represents here the pleasures of falling, slipping, sliding, tripping, collapsing, losing one's balance in enjoyment and torment: the only way to protect oneself from slipping is to simply sit still, sit well. This is the opposition between the faltering step and the sure seat, the faux pas and the *bienséant*. The princess trips on her dress and produces a lapsus, which sets a whole game of signifiers into play. Her fast recovery and her use of the misstep or faux pas as a pretext for retreat—"feignant de ne se pouvoir soustenir, elle s'en alla chez elle"—demonstrates how quick on her feet the princess can actually be when it comes to transforming a misstep into an acceptable reason, in the order of *bienséance*, to disappear entirely from the scene, to flee.[47]

Madame de Clèves's faux pas brings up the play of doubled interiority (she wants to stay, she wants to leave) and double dissimulation, one directed at the outside world and one directed within (she is trying to hide her emotions from the Reine Dauphine's entourage, and she is trying to hide her feelings from herself at the same time). "Elle s'embarrassa dans sa robe" means that she becomes entangled in the conflicted field of her desires. This kind of self-betrayal leads Jean Rousset to point out in his reading of the novel that it is actually on the scene of external life that truth about the inner life can be revealed.[48] It is only the world of masks and in theatricality of public life at court that the truth can betray itself. The princess contemplates what has been revealed to her in solitude, seated rather than standing, when she is well protected from the dangers of precipitation: "She locked herself in her rooms."[49] Her isolation gives her time to master her inner conflicts: all these locked spaces and secured interior places are the site of invisible and unrepresentable power struggles.

The faux pas arises out of the *faut pas*—"il ne faut pas aimer Nemours," and so forth—that is transmitted by the mother's lessons. Nemours is also torn between coming and going. During the second scene of his spying on the princess in her *pavillon,* he takes a few steps that, while bringing him closer to her, sends her into inaccessibility. After having observed the princess in her reverie with her ribbons, Indian cane, and paintings, he tries to approach her by entering the *pavillon* where the princess has been daydreaming: "Moved by the desire to speak to her and reassured by the hopes that the scene gave him, he advanced a few steps, but so clumsily that his scarf got caught in the window and he made a noise."[50] The sound of the scarf caught in the window alerts the princess to what Michel Butor describes as a "fantôme de la réalité," or ghost of reality.[51] Nemours's scarf gets caught in the window and this repeats the earlier entanglement of the princess in her dress. Accessories and fabric seem to get in the way of these two lovers with great regularity. The princess, thinking she has seen him, or his ghost, quickly withdraws to another room, where she is no longer in danger of being haunted by the phantoms of her real desire.

To compare the fates of the Princess of Clèves with her fictional cohorts, the Princess of Montpensier and the Countess of Tende, is to be confronted with the tragic consequences of the double bind of passion and renunciation. If the Princess of Montpensier and the Countess of Tende are punished for giving into passion, for disobeying the "il ne faut pas," they also enjoy an adulterous embrace. The Princess of Clèves is not in danger of adultery at the end of her story. She uses *bienséance* itself to dissimulate her dissimulation of the passion that Nemours has inspired in her. She extends the conventional time of mourning for a widow, after it seems the period of mourning for a husband might be over, in order to work out and work through her feelings for Nemours.[52] This peculiar strategy of mourning was already described by Madame de Chartres in her long discourse on the history of the Duchess of Valentinois. At the end of the story about the conflict between Madame d'Étampes and the Duchess of Valentinois, Madame de Chartres adds an enigmatic footnote on the death of the Duc d'Orléans:

> The Duke of Orléans died at Farmoutier of a contagious disease. He loved and was loved by one of the most beautiful women of the court. I will not name her because she hid with such great care the passion that she had for this prince that she deserves to have her reputation protected. By pure accident, she received the news of the death of her husband on the same day that she learned of the death of the Duke of Orléans; thus she had a pretext for concealing her real affliction, without having to suffer from self-constraint.[53]

This anecdote on dissimulation and mourning transforms the very possibility of reading the authenticity of grief. The death of the lady's husband gives her a convenient pretext for appearing aggrieved; her grief as a widow has been scripted in the codes of *bienséance* and is therefore readable, legible, and acceptable. The discreet and passionate lady hides a false grief behind a mask of authentic mourning. One loss hides the significance of another death: the husband's death requires a period of publicly expressed mourning and affliction but inspires little emotion. The death of the lover, on the other hand, causes great grief, which under ordinary circumstances would have to be hidden. The suffering from having to hide the grief over a lover's death is alleviated by a socially acceptable phase of mourning for a husband. The coincidental timing of the death reports provides the luxury of not having to hide her "real" affliction. This nameless woman and the Princess of Clèves both use the rules of *bienséance* governing the comportment of a widow in order to protect and dissimulate the marks of their own passions. In the case of the Duke of Orléans's lover, the heart and the countenance do correspond, but a level of deception persists: grief is expressed, made external, but this very display is a deception.

Madame de Chartres is the storyteller, whose disappearance paves the way for the novel: in her stories, wisdom of the generations is transmitted to her daughter in order to be of help, to provide counsel. This is what distinguishes a story from a novel, which according to Walter Benjamin does "not contain the slightest scintilla of wisdom."[54] That this novel represents the proper application of a storyteller's counsel does not necessarily make it the purveyor of wisdom. *The Princess of Clèves* leaves the reader with an inimitable example of enigmatic virtue.

The princess prolongs the socially required period of mourning for her husband in order to collect herself, to gather herself together to respond to Nemours. She is able to accomplish a *performance* of austerity in her gesture of refusal, and it is through this gesture that she is eventually able to immolate her passion. This is her great achievement of distinction: her renunciation becomes her glory. According to Madame de Chartres, introducing her daughter to the intrigues of life at court, "If you judge by appearances in this place, . . . you will be often fooled; that which appears is almost never the truth."[55] In this place, this world of deceptive appearances, there are many different ways to deceive and be deceived. Her warning is one that the reader of this novel cannot and should not fail to take into account: what is this place "ce lieu-ci," the court that offers a code of ritualized representations of private life? This is also the space of representation, of writing, and of the novel itself; it is the space in which a loss is always registered, even if it is under a different name. In the theatricalized world of the subject of courtly life, mourning is ritual, convention is the public commemoration of something lost. What must be mourned is the place in which appearances can be trusted and the sincerity of men remains unquestioned. This work of mourning is precisely what is avoided when we read the princess's decision as one that affirms sincerity. Something is lost when one is initiated into the ways of *this place:* the court marks and disfigures its members from whom terrible renunciations are demanded. For Howard Caygill, as for Samuel Weber, Benjamin's work on the German mourning-play isolates a historical place:

> For the Trauerspiel the world was empty, a place of "never-ending repetition" with no possibility of ever become genuine or authentic, "For those who looked deeper saw the scene of their existence as a rubbish heap of partial, inauthentic actions" (p. 139). The world handed down to us by tradition is uncanny, undecipherable, always other. History becomes an allegory, withholding its meaning just as it seems to offer it. Benjamin's reading of traditions stays with its destructive aspect: instead of authenticity within tradition, in a tragic fulfillment within time, tradition itself is inauthentic.[56]

When the feminist literary critics inscribe the princess as tragic heroine, they are repeating Coulet's reading of her as an icon of authenticity:

what DeJean, Miller, and Coulet have in common are modes of reading that isolate her from the court itself. Benjamin describes with startling accuracy the princess's own assessment of her "partial, inauthentic actions." Her refusal of Nemours's suit is founded on her inability to entrust herself to his sincerity. The "world handed down" to the princess is the one marked by a maternal interdiction: it is her mother who destroys her trust in appearances and in declarations of love.

What is masked by appearances is not necessarily a stable substantive truth but, rather, one that can be best described as Nietzschean—that is, will-to-power. Beneath the appearance of an austere and inimitable virtue, the princess can be said to be masking a will-to-power, deformed by the conventions of representation as well as by the codes of (feminine) comportment. Representation disfigures, but what can be retrieved of an original figure is figuration itself. The figure that reappears at the end of this novel is one that represents will-to-power. This will is so powerful that it blinds the reader just as it immolates the princess's passion. The truth of will-to-power is almost always masked, but it is nevertheless almost always legible. In this novel, will-to-power veils itself as inimitable virtue, but will is coded in the logic of superlatives that surrounds the character of the Princess of Clèves. The princess has achieved a radical singularity by the end of the novel, and this singularity functions like a signature. Authorship itself is allegorized here: this allegory of signing anticipates a set of problems investigated by Jacques Derrida in his work on the signature as both "absolute singularity" and "enigmatic originality."[57] Lafayette's refusal to sign *The Princess of Clèves* marks this novel and its readings with a distinctive absence. The author's signature is consummated, however, when the Princess of Clèves disappears into her own inimitability, when her proper name coincides with the impossibility of her representation.

This novel represents the problems of figuring and disfiguring will-to-power. We imagine that novels, like machines, work to hold something back from us. This pushes us, as readers, further and further into explanatory dilemmas. If the secret the novel seeks to protect has to do with an attempt to conceal its representations of thwarted ambition and unfulfilled love, it manages to do so by telling us a compelling story that poses a series of unanswerable questions: did Nemours really love her? Does it matter? It becomes obvious to us that the princess loves him but finds a way of becoming indifferent to him. That presentation within the novel, of an undisturbed surface of feminine virtue and stoicism, is based on her self-mastery. Refusing to be drawn into the disorders of passion, the princess discovers secret powers of self-control that allow her to achieve what Samuel Weber has called a "singular extreme."[58]

If passion is the realm of male control, then she has indeed escaped something, but what she achieves in so doing is a compelling distinction, an originary renunciation, whose consequences haunt many literary heroines for centuries to come. She has made history. It is difficult to think of renunciation as an authentically subversive gesture, but Lafayette does not pose its alternative as liberating. The author narrativizes self-discipline but does not penetrate its secret. The princess offers us a shining example of virtuous widowhood. Her disappearance from court and the unhappy outcome of her marriage and her love affair cannot be considered significant political events: it is in the uncanny exemplarity of her self-mastery that she distinguishes herself. To think of the classical novel as a defensive machine, a fortress, or a model of tyrannical mastery allows the contemporary critic to act as the one who reveals secrets and unveils machinations and mechanisms. It might lead us to believe that there are novels that offer themselves up without resistance to understanding; it might lead us to think that we have mastered something in the work itself. I would argue, rather, that the princess's secrets cannot be forced: she is not a figure with whom we can identify.[59] In trying to see ourselves in her, either as agents of decision or victims of "male control," we allow ourselves to be seduced by the glassy gaze of impassivity. Like Nathanael, we are bewitched by our own dilemma: we are trying to attribute human qualities to a literary character. To read a novel is to be seduced into identification: to reread a novel is to offer oneself the opportunity to resist primal identifications. The machine of surfaces presents us with many reflections of ourselves: it is in this specularity that we can be so easily caught in identificatory modes of reading.

The princess emerges as a force itself, the devastating force of self-discipline and will-to-power. The novel takes its shape from this force: we can recognize in it the principle by which every member of court submitted to the need to keep up appearances. That the constraints on women in both love and writing are greater is an incontrovertible condition of the seventeenth-century scene: that is why their aristocratic stoicism appears to be so moving, so radical, and so compelling. The princess was a harbinger of a modernity that would recognize itself in her insofar as she was the most complex and most interiorized of literary characters that had yet taken the scene. She also emerges as the tragic heroine for an entire class of ambitious aristocrats, who seeing their power diminished daily in the tyrant's court consoled themselves with representations of heroic renunciations.

Benjamin describes the advent of the novelist as being determined by the withdrawal of the novelist, who is faced with the incommunicability of modern experience: storytelling loses its force under such conditions. Lafayette's novel represents the retreat of the novelist: "The birthplace of the novel is the solitary individual, who is no longer able to express himself by giving examples of his most important concerns, is himself uncounseled, and cannot counsel others."[60] The silence and isolation of the princess are first and foremost literary. The perplexity and intensity of her inner life produce a new form of fiction. Ironically, the novel form, in which her experience finds itself most successfully represented, would evolve in the following century into the site of a new and highly mediated gregariousness, which while haunted by the austerity of her example would nevertheless attempt to rewrite the force of both her decision and her passivity.

Chapter 4
Getting Ahead with Machines?
The Cases of Jacques Vaucanson
and Thérèse des Hayes

Appearances

In Benjamin's "Theses on the Philosophy of History," the automaton-puppet dressed in Turkish attire wins every chess match against its human opponents. Inside the contraption, hidden by a system of mirrors, is a hunchback who happens to be an expert chess player, guiding the puppet by means of strings. This device is an ironic and ambivalent image of Benjamin's own methodology:

> One can imagine a philosophical counterpart to this device. The puppet called "historical materialism" is to win all the time against historicism. It can easily be a match for anyone if it enlists the services of theology, which today, as we know, is wizened and must be kept out of sight.[1]

The philosophical counterpart wins every game against historicism, but not because it cheats, although there is a trick involved. Its victories are scored legitimately, but theology, the wizened dwarf, must use an automaton dressed in fantastic Turkish attire as its proxy and mediator in the game it plays against historicism. The entire contraption—dwarf, mirrors, and automaton—comprises the complex figure of historical materialism. Benjamin implies that a historical materialist approach is dependent on a theological understanding of messianic time. In order to win the struggle for historical interpretation, Benjamin finds in historicism an adversary that must be confronted in an indirect manner. Historical materialism, or Benjamin's automaton, defeats historicism, but historicism can always accuse Benjamin of having played a cheap trick and hiding the wizened hunchback—theology, the secret of its success—from view. Is not the wizened, hunchbacked dwarf a legitimately better chess player than its adversaries if it can beat them at a game of chess? So what if the hookah-smoking automaton appears to be deciding? Theology may need to masquerade as an Orientalized automaton called historical materialism, but it scores legitimate victories, at least within the limited field of the rules of chess, which until now do not have anything to say about the appearance of the players.

The truth perhaps depends on a trick, and in order to win against historicism's notions of progress and continuity, one has to play a game with mirrors. In the struggle for representations of the past, Benjamin's automaton wins every time against historicism because the latter's simpleminded attempts at temporal homogenization tend to ignore the terrifying and destructive power of Messianic temporality:[2]

> Historicism rightly culminates in universal history.... Universal history has no theoretical armature. Its method is additive; it musters a mass of data to fill the homogeneous, empty time. Materialistic historiography, on the other hand, is based on a constructive principle. Thinking involves not only the flow of thoughts, but their arrest as well. Where thinking suddenly stops in a configuration pregnant with tensions, it gives that configuration a shock, by which it crystallizes into a monad.[3]

The additive method of mustering data to fill in empty time is based on a notion of progress whose political consequences produce the fatal compromises of Germany's Social Democrats. The automaton is a monadic figure, who represents both technological optimism and an uncanny, demonic double, whose imagined inauthenticity allows for the infinite deferral of a confrontation with thinking. The automatons of the ancien régime were playthings of the aristocracy, but by the nineteenth century they had become familiar and disturbing traveling sideshows of shady provenance. The automaton is slightly disreputable and untrustworthy. We saw the necessary denigration of the automaton, as an anthropomorphized machine in the work of Descartes; in the next century, this machine was to play two roles, one in the critique of worldliness, and the other in the exaltation of progress. Julien Offray de La Mettrie offers a critique of these two Enlightenment conceptions of the machine.[4]

Historians of philosophy like Ann Thomson warn against taking La Mettrie's conjugation of "Man and Machine" too literally. She argues that La Mettrie was not proposing to either construct a mechanical man or take apart the human organism like a machine.[5] Although La Mettrie does not go into great detail in his description of mechanical man, he uses mechanism as a speculative metaphor for complexity. La Mettrie saw the attempts of eighteenth-century engineers like Jacques Vaucanson to construct replicas of the human and animal organisms as compatible with his own project. In fact, La Mettrie's uses a comparison of Vaucanson's automatons to illustrate the different levels of complexity that can exist when comparing the human organism with simpler organisms like animals:

> If Vaucanson needed more skill to make his *Flute-Player* than to make his *Duck,* then it would be an even greater challenge to make a *Talking Automaton.* This machine can no longer be regarded as impossible, especially in the hands of a new Prometheus; . . . the human body is an immense clock that has been constructed with so much artifice and skill that if the wheel that serves to mark the seconds has come to a stop, the one that marks the minutes continues its course.[6]

That Vaucanson succeeded so well in creating mechanical wonders that imitated human and animal organisms seems to prove La Mettrie's point that organisms are governed by very complex, scientific laws that are analogous to the laws of mechanics. Vaucanson's success proves, above all, that while different levels of complexity determine the difference between animals and human beings, animal and human organisms exist nevertheless on a continuum.

L'Homme-Machine (Man-Machine) is primarily preoccupied with dispelling metaphysical notions of the relationship between soul and body by means of a discussion of the involuntary functioning of the nervous system. La Mettrie's insight into involuntary functions is garnered from very basic biological research that was being done in the eighteenth century by scientists like Albrecht von Haller, to whom *L'Homme-Machine* is dedicated. Haller was not appreciative of being cited in highly flattering terms in a text considered entirely heretical, but we have an example of the mischief that La Mettrie did not hesitate to make. (He knew Haller to be a devout man; he also sincerely admired his work.) He wanted to show that the refusal to compare man to animal/machine is a refusal of an analogical relationship, is pious, and has no basis as empirical evidence. La Mettrie tries to prove by means of involuntary movement that the body can function in the absence of the "soul" and that there must be something inherent in the anatomy of the organism itself that causes its movements. For La Mettrie, it was necessary to understand that scientists could study the organism, the source of its movements and ailments, and in doing so take it apart like a very complicated machine. The refusal to compare men and machine has recourse to absolute difference, while the bringing together of man and machine takes place as a rhetorical gesture of comparison. Man can be *like* a machine and a machine can be *like* a man. In this kind of comparison, a relationship of analogical rather than absolute difference is established between what man (or human) is from what he is not. By using the machine as a model of the human, La Mettrie establishes an analogy between organism and machine in order to theorize, justify, and call for the continuation of scientific investigation; further, he disposes of the question of Cartesian dualism by insisting on the continuity—that is, the undifferentiability—of *res cogitans* and *res extensa*.

The machine as a mediating double permits man to forge a different relationship with knowledge and speculation about the human body. La Mettrie set out to found the representation of interiority in anatomy rather than metaphysics. This interiority was the space that La Mettrie sought to represent by means of an image of immensely complex machinery. In this passage from *Man-Machine,* La Mettrie argues his point with characteristic irony:

> Experimentation and observation should be our only guides here. There are countless experiments to be found in the Records of Doctors who were Philosophers, but not with the Philosophers who were not Doctors. The latter have taken the measure of and explained the Labyrinth of Man. They are the only ones who have uncovered the springs hidden under envelopes that have kept so many wonders out of sight.[7]

La Mettrie continues to argue against the fanaticism of theologians who, without any kind of experience or observation, try to speculate on the nature of the human in total ignorance of the body's mechanisms. Doctors who are also philosophers are the best guides in the labyrinth of the organism; theologians, or fanatics, have gotten lost in it. This labyrinthine image of the interior of the body illustrates its awe-inspiring complexity, its secrecy and obscurity. If the interior of the body is a labyrinth, however, it is one that can be mastered by the *médecin/philosophe*. This labyrinth-machine that is man is "constructed in such a way that it is impossible to give a clear idea of it, and as a consequence, to define such an idea."[8] La Mettrie continues to use images of enormous complexity in order to illustrate the relationship between the doctor/philosopher and the space he sets out to explore.

La Mettrie argues that the nonsupremacy of the will is a result of the inseparability of *corps* and *esprit*. The body is subject to a myriad of conditions that also have a direct and powerful effect on the will of the mind:

> It is futile to make a fuss about the empire of the Will. For every order the Will gives, it is forced into submission a hundred times. It is a miracle that a healthy body obeys, because a torrent of blood and spirits are there to force it; the Will has as its Ministers an invisible, fluid legion that are quicker than Lightning and always at its service! But because its power is exercised by the Nerves, it is also by them that the Will is impeded. Can the greatest willingness and the most violent desires of an exhausted lover compensate him for his lost vigor? Alas, no; and the Will is the first to be punished because, under other circumstances, it is not in its power to not want pleasure.[9]

The failure of the will is best proven by the example of impotence, a biological failure, a disjunction between what the mind wants and what the body performs. Yet this disjunction is evidence of the way in which the will must submit to the "yoke" of bodily influences. The will has as its "ministers" a legion of invisible fluids, but its dominion does not extend to the nerves, which will defy it. The will here is personified as a king whose omnipotence turns into impotence when he is defied.

In the field of literary criticism, Benjaminian destruction takes place in one limited sense, as the undermining of differences between the order of truth and the logic of appearances. Benjamin's ironic use of the chess-playing automaton as a figure for historical materialism is grounded in this disruption. In Françoise de Graffigny's novel *Lettres d'une péruvienne* (published in 1747, the same year as La Mettrie's *Man-Machine*), we find a fully sentimentalized heroine, Zilia, a Peruvian princess whose critique of worldliness relies on a striking description of a doll/automaton.[10] Zilia has the self-assuredness of an aristocrat and the moral outlook of a bourgeois. Her confidence and steadfastness can be attributed to her high rank among Peruvians: but her critique of Parisian society is firmly rooted in the outsider's view, a position that Rousseau would elaborate on with even greater vehemence. As a princess and an outsider, Zilia represents the bourgeois subject's sense of superiority and marginality. Jean-Jacques Courtine and Claudine Haroche remind us that in the eighteenth century French aristocratic behavior had already been greatly influenced by a bourgeois sensibility, and that this influence mostly manifested itself in the eighteenth-century cult of sentimentality.[11] Aristocratic worldliness would be subject to ever harsher attacks throughout the century, culminating in the critique to end all critiques, the Revolution itself.

Zilia's observations about Parisian society culminate in the following description. The French are doll-like beings with an impoverished inner life; they act in a completely externally motivated manner. (By "the French" here, we are meant to understand the worldly aristocrats that Zilia meets through Déterville, the man who saves her from the Spanish conquistadors, and his family.) The implied objects of critique are society women:

> They are almost exactly like certain of their childhood toys: they are formless imitations of thinking beings. To the eyes, they seem to have weight, but they are light to the touch; they have colorful surfaces, formless interiors, apparent worth, but no real value. Other nations respect them as much as one, in society, respects pretty baubles. Happy is the nation that only has nature for a guide, truth as a principle, and virtue as the prime mover.[12]

The French have perfected the art of not appearing to be what they are: this description of the falseness of social beings echoes Madame de Chartres's warnings to her daughter about life at court, but the difference between the moral landscapes of *The Princess of Clèves* and Graffigny's novel is enormous. There is a happy nation over which "nature," "truth," and "virtue" rule: exotic Peru. Paris is the site of falseness, deception, and inauthenticity; Graffigny's fiction of the virtuous outsider relies on the Enlightenment trope of defamiliarization. From this place of natural virtue, the members of Parisian society will appear as imitations of thinking beings: their appearance is pure deception. The doll and the marionette as variations on the automaton become figures of inauthenticity and, specifically here, the deficiencies and dissimulations of worldliness. These doll-like beings manage to look heavy and substantial to the eye but are light when actually handled: they are truly virtuosic objects of deception. Their surfaces are seductive, but their interiors are hopelessly deficient in both form and content.

Graffigny's critique of the social forms of worldliness ironically inherits a great deal from seventeenth-century writers like François de Salignac de Fénelon who wanted a moral reform of the aristocracy in order to close its ranks to those bourgeois who imitated the manners of the aristocracy and were admitted into the salons.[13] Carolyn Lougee in her historical study of the significance of the debates concerning women in and around the seventeenth-century salon, *Le Paradis des Femmes,* shows how it was antifeminists like Fénelon and François d'Aubignac who wanted to stop the mixing of bourgeois and aristocrats in the salons of the *précieuses*. According to Lougee, in seventeenth-century France, women were the dominant arbiters of taste in the salons and determined the parameters of social refinement. She argues that those who were against the mixing of classes were against the worldliness of the *précieuses* and accused them of moral deficiencies: "There was widespread agreement that the salons were merely elegant brothels. Women bartered their bodies for social advancement."[14] The salons, then, were scandalous places where women could act on erotic and ambitious desires, and where refined and ambitious bourgeois mixed with nobles. Graffigny would agree with Fénelon on the moral depravity of the worldly environment: her proposed reformation would be based on a credo of virtue in turn based on sensibility and sentimentality. When we see Zilia's critique of worldly society in the context of seventeenth-century predecessors, we come to understand that Graffigny has set up her heroine as a bourgeois critic of unethical worldliness and that her "solid virtue" was based on bourgeois principles. On the other hand, Déterville's moral steadfastness has to do with the fact that he seems to be a member of the oldest part of the French aristocracy, the *noblesse d'épée*, which had seen its power progressively undermined during the changes that took place in the constitution of the aristocracy under Louis XIV's regime of ennoblements. Déterville is a soldier, not a courtier; he is a man of action and integrity. Graffigny's fictional attribution of bourgeois virtues to the members of the nobility of the sword is actually consistent with the most conservative of seventeenth-century views on social hierarchy.

The seventeenth-century critiques of worldliness that influenced Graffigny's perspective are for class stratification and antifeminism, that is, antisalon. For Graffigny, secure feminine virtue, however, does not lie in the realm of marriage. Zilia will not run a salon, nor will she get married; she has no worldly ambition, but she will be the arbiter of behavior and taste in her household. She will preside over her dominion in much the same way as Rousseau's Julie presides over Clarens. As Elizabeth MacArthur demonstrates in her essay "Devious Narratives: Refusal of Closure in Two Eighteenth-Century Epistolary Novels," marriage can be understood as a narrative device, and one that offers quick answers to all sorts of questions.[15] The *précieuses*, according to Lougee, were for marriages of convenience and extramarital affairs, as well as an active life in society for intelligent, engaging aristocratic women; the antifeminists, bourgeois and noble alike, saw marriage as the only destination of the virtuous woman. Graffigny's Zilia tries to avoid both society and marriage; this eighteenth-century variation on feminine destiny offers the heroine the company of her suitor, Déterville, but as a sublimated lover, a friend, in a situation that is designed to safeguard both of them from the passion of jealousy, a passion based on the exchangeability of one love for another. The Princess of Clèves leads a life so saturated with dissimulations and simulations that in the end the truth of her desire seems to become the promulgation of pure appearance. Fiction relies on literary devices; identifying them, however, does not necessarily defuse their gadget-like power to produce unexpected effects. The princess's decision remains as mysterious as ever: the fact that Graffigny's Zilia also refuses love, in the name of the cultivation of virtue, is founded on the novel's establishment of her happiness in social retreat and peaceful friendship with her suitor. This rewriting of feminine refusal tries to ground virtue in the space of authenticity, but as we saw with Julie, Rousseau's attempts to destroy the difference between the internalized secrets of the heart and the externalized countenance produce the imposition of an absolute authority. Wolmar's utopia of sincerity is a difficult fiction to sustain: even in Julie's carefully laid-out space of domesticity, things are not always what they seem. It has been argued that the disjunction between truths and the appearances drives a narrative forward:[16] once that distance has been abolished, there are no more stories to be told. Only a catastrophe can shatter the static calm of such transparency.

In the examination of Jacques Vaucanson's career as one of the most famous eighteenth-century French automaton makers, we find a relationship between the literary and philosophical figure of the automaton, and the historical circumstances around the biography of this minor Enlightenment figure. In interrupting the work of literary analysis, we attempt to follow the very form of Benjamin's thinking, described in the following manner by Irving Wohlfarth:

> As it moves between theology and historical materialism, and establishes ("illegitimate," "free-floating"?) relations between Moscow, Berlin West and Jerusalem, Benjamin's thinking suddenly stands still amidst a force-field—it was increasingly turning into a mine-field—of tensions and crystallizes around a figure *(Denkbild)* which leaves none of its points of reference intact.[17]

This figure around whom I have constructed a number of relationships and associations is the automaton.

Vaucanson, according to his biographers A. Doyon and L. Liaigre, is supposed to have cherished the dream of creating perfect copies of anatomical and organic processes through his mastery of the laws of mechanics.[18] He made a few tricky automatons of his own, and his famous automaton-duck was hailed as one of his great successes. He makes it clear that his intention in constructing this particular automaton was to prove that mechanics was entirely capable of imitating and accurately re-creating anatomy and anatomical movement. The automaton-duck was part of the debate among doctors, anatomists, and surgeons on the nature of digestion. Vaucanson was opposed (ironically enough) to a purely mechanistic model of digestion then popular among powerful doctors like Hecquet, a court physician protected by the Prince de Condé and former doctor of the Port-Royal who had published *Traité de la digestion et des maladies de l'estomac suivant le système de la trituration et du broiement* in 1711.[19] Vaucanson became what Doyon and Liaigre call, perhaps hyperbolically, a "theorist of digestion" by criticizing Hecquet's description of digestion as a process during which food in the stomach was simply pulverized by mechanical motion. He enters the heated debate and confronts one of the Parisian medical establishment's most influential figures in denying that digestion involved trituration, or tiny, imaginary saws and grinders that crushed food, reducing it thereby to a nourishing pulp.

Doyon and Liaigre are eager to cast Vaucanson in a heroic role in this conflict. Vaucanson is the outsider who speaks out in the name of truth against a powerful man. In his letter to the Abbé Desfontaines, he describes in some detail the functioning of his duck (as well as two other automatons—a flute player and a figure playing a drum and fife). According to Vaucanson's description, the mechanical duck is able to extend its neck, flap its wings, and take food from a hand that feeds it, swallow the food, and through more neck movements pass the food into its stomach. He writes that in his duck, "as in real animals, the food is then digested there, through a process of dissolution and not by trituration, as many doctors claim. But this is what I will deal with and demonstrate in right time. In the stomach, the digested matter follows tubes, that are like the entrails of animals until it ends up in the anus where a sphincter allows for its expulsion."[20] Vaucanson explicitly claims that a process of digestion actually does take place. In order to defend his copy as an accurate but imperfect copy, he limits his imitation to three aspects of duck movement and duck anatomy:

> I do not claim to offer this digestion as a perfect digestion, capable of producing blood and nourishing particles for the survival of the animal;

one would be in bad faith in reproaching me for this. I only claim to imitate the mechanics of this action in three ways, which are (1) the swallowing of the grain; (2) the maceration, the cooking, or the dissolving of the grain; (3) the expulsion of the grain in a visibly altered form.[21]

The problem was that this controversial automaton-duck depended on what Doyon and Liaigre call "une supercherie," a sleight of hand, or what others might want to call a bit of cheating. The excremental material the automaton-duck succeeded in expelling had nothing to do with the seeds it had swallowed; these pellets were prepared in advance and located in a hidden container in the posterior of the mechanical animal.[22] Thus while Vaucanson's model was flawed, it was nonetheless closer to our understanding of digestion than the one that it sought to disprove.

History

For Benjamin, history can be read for moments of crystallization because historical time is not the time of linearity and progress; it is, rather, the time of repetition and allegory. To read repetition in history is to read history as containing within it hard, shiny kernels of the present, the present that consistently escapes us. Benjamin criticizes historicism in the name of historical materialism because of historicism's inability to read the present in the past: historicism's linear notion of time and its construction of events as isolated singularities on the march of progress inherit the Enlightenment's own construction of time, space, and knowledge. Biography is an interesting case of the writing of a kind of history. Doyon and Liaigre's biography of Vaucanson, *Jacques Vaucanson, mécanicien de génie*, tells the story of a man who was ruled by the ambition to perfect mechanical copies of organic originals, that is, automatons. He finds himself at the end of life, serving commerce and the king in improving mass-production techniques in the Lyon textile industry. The biography not only tells the story of Vaucanson's life: it also argues for the fact that the technology of the automaton gave way to the principle of automation. The automaton is a figure of both repetition and allegory, of the radically discontinuous temporal relationship that connects us and cuts us off from the origins of modernity.

Vaucanson's automaton-duck and Benjamin's automaton of historical materialism share the fact that each contraption contains a trick, but that despite this trick, these automatons have a better claim on representing truth. In *L'Institution de la science et l'expérience du vivant*, Claire Salomon-Bayet credits the mechanistic theory of the eighteenth century for providing science with the living being as an object and a model by means of metaphor and symbol:

> It seems as if mechanics as a system was doomed to failure in the context of life science because it made experimentation impossible.... This failure, however, was also a success: mechanistic theory, because it took the model, the symbol and the metaphor as its objects, made possible the constitution of the living as a positive, experimental and limited science.[23]

His failure is his greatest success: for in attempting to reproduce anatomy through mechanics, Vaucanson actually anticipates a relationship, mediated by symbolization and metaphor, between modern biology and its object—life. It is in this particular objectification of the living that experimentation is rendered conceivable. It is only *après coup* that Vaucanson's mechanical legerdemain can make sense. According to Jean-Claude Beaune, "It's when Vaucanson is cheating that he is the most *savant*.... The automaton, having become an experimental ruse, nourishes... a new spirit of inquiry."[24] Even when deceit is involved, the spirit of inquiry can be nourished by a mechanical duck.

The famous Vaucanson automatons of the eighteenth century exist only as images and descriptions now: it has been a century since they were destroyed. Unlike the machines of mass production, they defy reproducibility. Their unreproducibility makes them singular machines: their inimitability resembles that of Lafayette's mysterious princess, as an inimitable model or as an impossible signature. Like the princess who does not die but, rather, disappears into her inimitable singularity, many of the ancien régime automatons from France meet the same fate: this disappearing act is a literary trick, that is, a literary *device* that can serve to deceive. The device divides literature from any purely utilitarian use of language: wizened dwarves and secret compartments are evocative but inadequate as images in describing the duplicity of and pleasure in reading literature. In failing to be straightforward, the device succeeds in being literary.

Jacques Vaucanson's success story is a narrative that repeats and rehearses the narrative of countless literary arrivals: a gifted provincial arrives in Paris, is initiated into the secrets of its inner circles. This *mécanicien* of the eighteenth century rises above his class and gains entry into the spheres of privilege by virtue of his ingenious automatons. The life of Jacques

Vaucanson as a detour from and reentry into fiction should be considered, in order to be fully grasped, in the context of literary arrivals in ancien régime literature.

Vaucanson's automatons functioned simultaneously as objects of scientific inquiry and popular curiosity, although historians of science, when trying to establish his importance, like to emphasize the former at the expense of the latter. This complicitous relationship between science and curiosity is clarified by a consideration of the question of *Technik/technique*. *Technik/technique* describes both the skill of the craftsman and the bringing forth of objects into existence by means of his artisanal skill. Vaucanson's automatons crystallize a certain Enlightenment moment when the machine was not yet completely subordinated to the forces of industry. According to Beaune,

> The automaton that contains its own principle of movement radicalizes the geometrical machine, giving an almost infinite potential to the schematization of real movement. The lever-operated automatons of Vaucanson or Jaquet-Droz finally allow for the conceptualization of movements that follow one another and reproduce themselves infinitely, or almost infinitely. We have been able to see subsequently that it is also at this historical moment that the automatons seem to disappear. The juxtaposition of the knowledge of the body and automating techniques is invalidated, and is, in any case, no longer justifiable.[25]

It seems that by incarnating or schematizing the possibility of infinite, geometrically determined movement, the automaton as singular, mechanical object engineered its own obsolescence. The automaton of the seventeenth and eighteenth centuries is an "aristocratic" machine, singular and "precious" in every sense of the term and emblematic of the ancien régime. Vaucanson's career spans the passage from automaton to automation, from singular, autonomous machine as precious object of curiosity to mechanical principle and the interlocking, multiple machines of industry. Beaune describes this transitional period: "The automaton's analytic functions are displaced, and they continue without it: the automaton disappears as solitary model and object in order to determine forms of work in an industrial milieu."[26] Before we can conceive of industrialization or automation, we must be able to conceive of the notion that the movements of human limbs can be mechanically represented—precisely what Vaucanson achieves with his automatons with their articulated limbs, fingers, and tongues. The eighteenth-century automaton is an early technological object whose relationship with craftsmanship and artisanal skill is still apparent; it was in reaching the height of its complexity that it was poised to make good its disappearance.

Vaucanson

Exploring the secret of Jacques Vaucanson's success will shed light on the struggle for power between science and spectacle, nascent capitalism and the despotism of kings in mid–eighteenth-century Paris. The *Encylopédie*'s entries under *automaton* and *androïde* are devoted almost entirely to his work. Vaucanson's achievements were successfully woven into Enlightenment history. In 1738, Jacques Vaucanson, an obscure but gifted young provincial from a petit-bourgeois family of Grenoble, attracted the attention of the Parisian aristocracy and literati with his automaton, a flute player, which was capable of playing twelve airs on the *flüte traversière*. The flute player was based on a 1709 sculpture by Antoine Coysevox, *Berger jouant de la flute*. In the pedestal was lodged a wood cylinder, 56 cm in diameter and 86 cm long, which served as a "program" for the entire mechanism. The cylinder was carved to set off, while rotating, fifteen different levers that controlled the movements of the flute player's articulated tongue, fingers and the reservoirs of air.

In 1741, three short and eventful years later, after having obtained the recognition of the Académie Royale des Sciences through the judicious use of his worldly contacts and alliances, Vaucanson became, by command of the King Louis XV, "inspecteur des manufactures de soie." He was hired by the far-seeing Minister of Commerce Jean Orry as a government official and servant of industry at a salary of six thousand livres a year.[27] The French government was hoping to restructure silk production in Lyon and southeastern France in order to compete with the silk industry of Piedmont. Perhaps because his biographers Doyon and Liaigre wanted to compensate for the relative obscurity of their subject, they praise Vaucanson as a genius, as "the greatest French *mécanicien* of all time." Doyon and Liaigre do not fail to include the requisite anecdotes of inimitable childhood precocity and undaunted commitment in the face of adversity.

One way that we can begin to grasp the measure of Vaucanson's significance is to consider the fate of his home and workshop, the Hôtel de Mortagne, located on what is today the rue de Charonne, where most of his innovations in the silk textile industry were fabricated. After his death, Louis XVI bought the hotel from the Chevalier de Ham, Vaucanson's landlord, and declared the site on August 2, 1783, "public repository of models for machines used principally in the arts and in manufacturing."[28] The decree also stipulated that Vaucanson's collection of tooling machines eventually be supplemented by the purchase of the latest machines used in England and Holland (two countries that had outpaced France in industrial development and foreign trade—in short, in the major areas of capitalist activity), in order that French artisans might benefit from them. In addition, this public collection of machines was supposed to inspire capitalists to invest in the production of new machines: the opening of Vaucanson's atelier was intended to encourage French industrialization and capitalism.[29]

Vaucanson was born in Grenoble to a family of humble origins on February 24, 1709. His father, who died during his childhood, was a master glove maker. The young Vaucanson began his education with the Jesuits in Grenoble; little is known about his childhood. In 1725, we find that Vaucanson has been accepted as a novice with the Minimes of Lyon. There was a scandal involving automatons that he had constructed in secret, which the Superior ordered destroyed. Vaucanson was released from his vows but did not give up all ambitions of one day entering the Church, with the primary hope (according to Doyon and Liaigre) of elevating his social status.

After a few years spent in Paris (1728–1731) that remain largely undocumented, Vaucanson traveled the north and west of France, displaying his automatons for a living.[30] In Rouen, he may have met the soon-to-be-famous surgeon Le Cat, who was very interested in the construction of artificial human anatomies; it is perhaps here that Vaucanson studied human anatomy and first conceived of constructing his *anatomies mouvantes*. In 1732, during a stop at Tours, Vaucanson met Jean Colvée, a wealthy monk at the Collégiale Saint-Martin, who was inspired enough by the young man's talents to invest in his future. In 1733 he agreed to finance Vaucanson's next automaton with a sum of 2,400 pounds, which went toward his daily needs during the time of its construction.[31] Vaucanson would repay Colvée by giving him two-thirds of the profit, until the sum was paid off; afterward, the good monk would earn 20 percent of all other profits. If Vaucanson could not repay him, Colvée would take the machine as a deposit on the outstanding debt.[32] If more funds were needed for the automaton's completion, Vaucanson would have to raise them himself.

Colvée worked out a hard deal with Vaucanson, and it seemed that his investment would pay in one form or another because, as the capitalist in this couple, he controlled the means of production and had bought

Vaucanson's expertise and labor. For the impoverished young man, however, the infusion of 2,400 pounds meant that he had found a way to assure his livelihood and install himself in Paris with a considerable sum of money. He chose to live on the Left Bank, on the rue du Four, right next to the Foire Saint-Germain (Saint German market grounds), where he may have continued to display his automatons from the northern tours for a bit of money on the side. In an act of pure extravagance, hubris, and ambition, he rented as his studio one of the rooms of Hôtel de Longueville, rue St. Thomas du Louvre.

A little time later, Vaucanson became a guest of M. de la Poupelinière, a great patron of the arts and sciences and one of the greatest libertines of his time. Voltaire frequently visited La Poupelinière, as did Rameau and many other distinguished figures of the time. It was here that Vaucanson was initiated into the life of Parisian worldliness. It was a frivolous milieu; to a young man from the provinces, it must have appeared a magical place.[33] Colvée, worried about his investment, traveled to Paris to find that another 2,400 pounds were needed for the completion of the project. Vaucanson managed to extract more money from his investor. It was not until 1742, well after Vaucanson had achieved financial stability, that he was able to repay Colvée in full. According to his biographers, the belated payment of this debt reveals something about Vaucanson's character. They call it his bad faith. By 1735, he was already wearing a sword, dressed like a gentleman in floral jackets and living in grand style.[34]

Certainly, the distractions of Parisian society are not to be underestimated. From the very beginning, Vaucanson seems to have been an ambitious young man whose commitment to science and mechanics seemed to be determined by his financial needs and his desire for recognition. Doyon and Liaigre recount that it was only after falling gravely ill in 1735, from an anal fistula, that Vaucanson decided to construct the long-awaited automaton in order to remedy his near-disastrous financial situation. It seems that shortly after this decision, he left La Poupelinière's salon and was received in the home of Jean Marguin, a bourgeois of Paris, from whom he received the sum of 3,000 pounds in return for half of the profits from the automaton. Doyon and Liaigre depict this agreement as draconian and see in it the signs of Vaucanson's desperate need for money. Vaucanson had mortgaged all profits to his creditors. Before the completion of the flute player, Vaucanson would have to borrow another 3,000 pounds from Marguin. In February 1738, the flute player was completed. According to his biographers, Vaucanson showed his automaton at the Foire Saint-

Germain for fifty louis a day for eight days before the opening of the exposition at the Hôtel de Longueville without Marguin's knowledge in order to turn a tidy profit for himself.

On February 11, 1738, the first demonstration of the flutist took place at the Hôtel de Longueville. These performances turned out to be highly profitable.[35] The Abbé Desfontaines, who also frequented the table of La Poupelinière, became one of Vaucanson's greatest publicists and wrote about the flute player in highly laudatory terms in his journal *Observations sur les écrits modernes* on March 30, 1738. He writes that from a man as skilled as Vaucanson, there was nothing one could not expect.[36] The *Mercure de France* was more reserved in its appraisal of Vaucanson's achievement. In the years that followed Voltaire put his hyperbolic praise of Vaucanson into verse.[37] La Mettrie calls Vaucanson "un nouveau Prométhée," but he was also interested in the limitations of the mechanical model.[38]

Despite the automaton's success, Vaucanson did not pay Marguin's percentage regularly. As one can imagine, the young *mécanicien* began to realize that he had cut a bad deal. Marguin appealed to the court of Châtelet to name a trustee for the Marguin-Vaucanson association. Vaucanson, feeling the pressure of this legal procedure, turned to a higher authority. He sought the help of the king and his council in blocking Marguin's legal efforts against him. In his letter to the king, he describes himself:

> Jacques Vaucanson, having applied himself to the sciences since his youth, had spent all the slight fortune that he had from his father in this endeavor. It was in this state of exhaustion, feeling the impossibility of completing the moving anatomies that he had started, that he thought of producing a few machines that could inspire the curiosity of the public in order to earn some money. He conceived of the design of a statue that could play the flute with an embouchure by the movements of its fingers.[39]

Vaucanson says he has ruined himself in his commitment to science: anatomical models of purely scientific interest were simply too expensive to finance for a young man of little fortune. Aside from the automaton-duck, Vaucanson never produced another anatomical model. The promised "moving anatomies" remain virtual objects, representatives of an impossibly disinterested scientific investment. The ruinous expenditures demanded by science and the slimness of his paternal inheritance drive him to exploit or inspire the curiosity of the public with his automatons in order to finance his scientific pursuits. According to his letter to the king, the flute player was born out of the financial distress of a young man who would have liked to have been able to devote himself completely to science. According to Vaucanson, making automatons for profit and entertainment was merely a detour from this original intent.

According to Vaucanson's version of things, Marguin uses the arts and sciences as a cover-up for his real, venal motive—profit and exploitation. He attracts and seduces the young man into his home under false pretenses; Vaucanson describes himself as being manipulated into accepting "two illicit and unpleasant stipulatons" ("deux actes aussi illicites qu'onéreux"). These financial transactions take on a sexual tone. Only with His Majesty's help can Vaucanson hope to make himself useful to the public.[40] To sum up briefly: the financial weakness of a father does not allow the son to pursue science in a disinterested manner. The paternal lack feminizes him, puts him in a weak position with regard to other men. Thus, out of desperation and lack, he is seduced by Marguin into an unpleasant association. Now he must appeal to the king to protect him from the exploiter. The king will intervene as a strong paternal figure and save the talented son from his feminization at the hands of a venal impostor.

Heidegger reminds us, "Pure science, we proclaim, is 'disinterested.'"[41] Thus proclamations are made for science as "disinterested," but in the quotation marks surrounding the term, we find a certain amount of irony. In his characterization of the king as a real protector and patron of the arts and sciences, Vaucanson appeals to the presumed disinterestedness of science. He sets up Marguin, in contrast, as someone who pretends to care about the arts and sciences but in the end loves nothing but profiting from a young man's talents, naïveté, and financial distress. Only the king can reward talent and punish venality. In their article "The Motives of Jacques de Vaucanson" David M. Fryer and John C. Marshall seek not only to clear Vaucanson of any suspicion of financial motivation; they also endow the engineer with an aristocratic particle. They too are determined to clear Vaucanson of any suspicion of an intention to make money and entertain the public. For these historians, a scientific hero is immune from material needs and possesses purely scientific and theoretical ambitions; thus they participate fully in the myth of scientific disinterestedness.[42]

Louis XV approved a suspension of Marguin's case against Vaucanson. This decision bought precious time for Vaucanson. The judgment of the king's council sent the two plaintiffs to the lieutenant Hérault, Councillor of the State, also a regular of La Poupelinière's soirées. Marguin still hoped that ordinary jurisdiction would rule in his favor because he found all his agreements in order, while Vaucanson hoped for a special dispensation. The details of how Vaucanson was able to eventually reach new agreements with both Marguin and Colvée are too laborious to recount here. Under great pressure from above, both creditors renounced their rights to the automatons and any future profits thereof; both also stated that they would be satisfied with repayment with interest of the loans made to Vaucanson.

Royal intervention helps free the young man from his obligations to his exploitative bourgeois investors. It was on the grounds of being wholly devoted to science that the young Vaucanson appealed to the king and his council to rule in his favor. The young Vaucanson, however, benefits from an anticapitalist, feudal system of royal intervention, which discouraged the French bourgeoisie from speculating and investing in technical innovation like their British and Dutch counterparts. Bourgeois interests were not guaranteed under French law as long as royal intervention could void the terms of exploitative contracts.

After having disentangled himself from his investors, Vaucanson now focused on gaining the recognition of the Académie Royale des Sciences, which was not kindly disposed to the science of mechanics, passing as it was through a period of great idealism about abstract geometry. Yet by this time Vaucanson was in a social milieu that allowed him to dream of acceptance by this body. In April 1738, M. de Fleury, the head of the Académie, ordered members to attend a performance of the automatons at the Hôtel de Longueville. On April 30, Vaucanson presented the *Mémoire descriptif* of his flute-player automaton. Three days later, the secretary of the Académie at the time, Fontenelle, drew up the much-desired letter of approbation.[43] In 1746, he was accepted as a member of the Académie itself.

Exciting Curiosity

On September 8, 1739, Françoise de Graffigny, still fresh from the excitement of her recent arrival in the capital, describes the activities of the day to her longtime correspondent, Devaux (affectionately known as Pan Pan), who remained behind at the court of Lorraine:

> Yesterday morning, I visited the King's library with Md. D.C., his Swiss Guard and V. . . . I saw a great many manuscripts that had been gnawed by rats. A scholar would have been delighted, but they did nothing for me. That afternoon, we went to see the flute player, and it gave me great pleasure.[44]

The flute player she refers to without much explanation is of course Vaucanson's automaton. Graffigny does not hesitate to describe herself as a "nonsavant," but she does go on to note her pleasure in seeing the spectacle of the mechanical device.[45]

In Alfred Chapuis and Edouard Gélis's study *Le Monde des automates* (1928), the authors speculate on a mimetic drive that motivates human beings to pinch bits of clay into anthropomorphic forms and then make the clay imitate human gestures.[46] According to them, the articulated statues and toys of the Egyptians are related to the articulated masks of tribal cultures in which animism plays a dominant role in mediating relations between human beings and their surroundings and their dead. Automatons not only help the living with the work of mourning; they play a crucial role in a utopic world where labor becomes abstract (let us recall that forced labor is at the Czech root of the word *robot*). In Aristotle's *Politica*, we find the description of a universe where things serve and obey the will of human beings; Vulcan's tripods serve the gods, musical instruments play themselves, and entrepreneurs and masters would no longer need either workers or slaves, as things animate themselves to serve the will of the deities.[47]

The origins of automatons are also associated with the very origins of art itself. Chapuis writes, "We think that primitive, articulated figures were one of the first manifestations of art. Man, in imitating nature, tried to reproduce movement. This movement was pleasing to the eye (when it did not inspire fear), and the artificial representation of life became very early a form of popular amusement."[48] As an illustration of how this "artificial representation" became a source of popular entertainment, Chapuis refers to a photograph of a carving of a processional mask from Bali that was constructed with an articulated jaw. The automaton was an anthropological object that once gave pleasure and frightened the public. Displayed in a ritualistic manner and in chronological order, the articulated statues and masks of "primitive" cultures are the ground on which the construction of a purely rational relationship to technology and its history can take place. Historians of science can answer the questions of the automaton's genealogy with such unshakeable certainty only if they suppress the relationship between the recent history of the automaton in Enlightenment Europe and these "primitive" counterparts.[49] Vaucanson took advantage of the automaton's pleasure-producing potential; he exploited its status as object of curiosity, its simple entertainment value. Inspiring curiosity, however, at least according to Vaucanson's letter to the king, was not his original intention. Through the disinterested pursuit of science, he wanted to make himself useful. Curiosity, however, can turn out to give way to the useful. To make of science a diversion is to rehabilitate the much-maligned notion of curiosity. The automaton turns out to be a model for thinking and representing human anatomy,[50] but in Vaucanson's version, instrumentalization of the pleasures of curiosity justifies his determination to perfect his machine. He endows his model with not only a didactic power but also the potential to yield new discoveries in order to add to the storehouse of knowledge that the *Encyclopédie* already represented. The automaton/machine at the service of knowledge, then, comes to embody "the additive" principle described by Benjamin as historicism's method.

When Beaune argues that Vaucanson's achievement lay in the fact that he was able to transform the automaton-curiosity into the machine-tool, he also partakes in the continuing denigration of curiosity.[51] The utility of the "machine-outil" offers Vaucanson a more legitimate place in the history of technology and the progress of industry. Curiosity is merely a detour on the way to utility—not a destination in and of itself. In her preface to the 1985 republication of Vaucanson's memoir, originally presented before the Académie Royale des Sciences in 1738, Catherine Cardinal tries to account for Vaucanson's intentions in creating his automatons:

> In constructing automata, Jacques Vaucanson did not simply want to produce creations of an astonishing complexity that would inspire the curiosity of gawkers. His design was much more ambitious. With his mechanical talents and his knowledge of anatomy, he wanted to create artificial beings, "moving anatomies." These anatomies were to have reproduced as faithfully as possible the organs and the functions of the human being or the animal. Their real aim was not to amuse, but to instruct and in so doing to further the progress of medicine.[52]

Curiosity is described as a quality of gawkers, the *badauds* who stop and stare at the spectacles of the street, and whose fascination is founded on their ignorance. The real goal of Vaucanson's automatons is pedagogical—any form of pleasure that they might inspire is denigrated as marginal to the goals of scientific progress. In order to inscribe Vaucanson's automatons in a history of the progress of science, Cardinal must insist that they are not merely objects of curiosity. Cardinal attributes to the automatons a "true" scientific ambition. The artificial beings and the moving anatomies about which she writes, however, are never really produced, only alluded to: the problem is that Vaucanson's automatons were curiously useless at first glance, and at least according to Graffigny, purveyors of a great deal of pleasure.

The entry for *curiosity* in the *Encyclopédie* affirms, however, that it "is the desire that inspires man to extend his knowledge, either to elevate his spirit to the great truths, or to render himself useful to his fellows citizens."[53] In Jean d'Alembert's "Discours préliminaire" to the *Encyclopédie*, we find a defense of curiosity and the curious. In keeping with the *Encyclopédie*'s definition, d'Alembert tries to show that the curious can always become or give way to the useful: in fact, curiosity is supplementary to utility and necessity. Curiosity is on the side of amusement, astonishment, gawking, and diversion: it is what produces pleasure in excess of the knowledge of the useful and the necessary:

> The mind, which is used for meditation and is eager to derive from it some fruit, must find, then, a kind of resource in the discovery of exclusively curious properties of some bodies. This kind of discovery knows no limits. In fact, if a great number of pleasurable pieces of knowledge are consolation enough for having been deprived of a useful truth, one could say that in the study of Nature, when we are refused the necessary, we are at least provided a profusion of pleasures: it is a kind of excess that supplements, however imperfectly, that which we are lacking.[54]

The pleasures of curiosity, according to d'Alembert, compensate the mind for its failure to produce useful and necessary knowledge; in short, it is a kind of consolation prize to the *esprit* for having reached a limit of reflection. The curious, as pleasure-giving principle, serves to supplement that which we are lacking. Derrida's reading of the supplement in Rousseau opens the way to an understanding of the excessive nature of the nonessential and the secondary. The fruit of meditation is supposed to be a useful truth, but when thinking is unable to reach this goal, it can have recourse to the pleasures of the curious, useless, and inessential. As Derrida has shown, however, there is something dangerous about this excess.

D'Alembert's curiosity comes to the rescue when the thinker reaches an impasse. Curiosity as supplement insinuates itself into the empty space that opens up when thinking fails the thinker.[55] The discoveries made available by curiosity seem infinite as opposed to the useful truths, which are limited by nature and difficult to come by. In place of a single useful truth, a profusion of curious facts are provided by Nature for the thinker in order to console him in his frustration. The danger of curiosity lies in the fact that it can throw the thinker completely off the track of useful truths. Curiosity sidetracks meditation, but it becomes, strangely enough, essential to thinking and the thinker:

> In the order of our needs and the objects of our passions, pleasure occupies one of the first places, and curiosity is a need for those who know how to think, especially when this anxious desire is animated by a sort of frustration at not being able to entirely satisfy itself. Another motive serves to sustain us during such work: if utility is not its object, then it can at least be the pretext. If we have discovered, a few times, a real advantage in certain areas of knowledge, where at first we had not expected it, we can authorize ourselves to consider all investigations of pure curiosity as being potentially useful. This is the origin and the cause of progress in this vast science, called Physics, or the study of Nature, which is made up of so many different parts.[56]

Curiosity is now implied as the original necessity and the real motor of progress. It designates a necessity beyond necessity. It is an anxious desire that precedes all need. It is also a pleasure principle that supplements and sustains thinking when the thinker is unable to completely satisfy himself. A thinker can and must know how to gawk productively at useless knowledge. Therefore, curiosity produces something absolutely useful: the sustenance of thinking. When one is pursuing a line of thought out of curiosity, for the pure pleasure of it, utility becomes a pretext, but curiosity and pleasure are the subtext.[57] Because of the uncertainty of the outcome in investigations that seem to lead nowhere, progress in the natural sciences is made. To keep progress on the main road of usefulness, one must take the detours of curiosity, even though these paths may appear at first futile. This is why the thinker must allow himself to follow his curiosity when he cannot completely satisfy himself in making clear progress.[58] Curiosity produces more thinking when thinkers are stymied: this is why a thinker needs to be curious and not just utilitarian. In the natural sciences, thinking produces usefulness and progress, but what is most useful to progress is curiosity. To renounce curiosity is to renounce thinking.

Automation

Vaucanson was transformed from a maker of mechanical curiosities into a useful subject of the king when he accepted a position as "inspecteur royale des manufactures de soie." He made a number of improvements in French silk manufacture, and he was also zealous in performing administrative duties that precipitated a crisis in the textile industry itself. One of his first responsibilities was to impose governmental reforms, which led to the 1745 strikes of textile workers in Lyon—the greatest strikes that the ancien régime was to know. In his "Éloge de Vaucanson," Condorcet describes Vaucanson's accomplishments in silk production as exemplary of technical innovation in the service of industry.[59] In praising Vaucanson, Condorcet defines the innovative role of the engineer. What he should do is substitute machine operations for the exercise of human intelligence: this leads to an improvement in the speed of production and the quality of the product. Vaucanson's innovations not only increased the rate of production; they also set out to improve of the quality of French silk. His inventions, like the *moulin à organsiner* that produced silk organza, made it possible to increase the uniformity of the final product. Vaucanson's early assembly-line notions of production certainly contributed to the decline of the power of *maître-ouvriers* in the production of silk in Lyon.[60] The mass production of textiles required that each worker perform a simplified task on an early assembly line in larger mills. These are, of course, the assembly-line techniques of mass production that Marx would criticize in the next century for stripping workers of skill and the pleasure in exercising that skill.

Vaucanson and his partner, Montessuy, had to enforce unpopular administrative reforms of silk production that reestablished a hierarchy between the small producers and the *maîtres-marchands fabricants*. In 1737 the smaller producers had been able to gain a certain amount of autonomy; by 1744, they saw their hard-won rights taken away. The strikes began in August 1744; they were violent and violently suppressed. The leader, Marichauder, was condemned to be hanged in March 1745, but the king gave him amnesty the following day. The great strikes of Lyon did not, then, have anything to do with workers' resistance to Vaucanson's innovations: instead, they were a response to the enforcement of new regulations, unfavorable to the interest of small, independent silk producers, by the new "inspector" who endangered himself personally in order to carry out his task.[61] Vaucanson proved himself loyal to the will of the state; he was the perfect civil servant and courtier. He was never to complete a "moving anatomy" but spent the rest of his life in the service of the king. His reputation for being able to extract the highest fees for all his services was to be mentioned, albeit in laudatory terms, in Condorcet's official eulogy.

Condorcet's "Eulogy of Vaucanson" tells the story of an undeniable genius who cannot be impeded on his path toward recognition and approbation. In this version of Vaucanson's life, the young *mécanicien* makes seamless progress from Grenoble to Lyon to Paris. Condorcet is not a biographer, after all; he is writing a eulogy and performing the ritual of flattering the recently deceased. What is interesting, however, is that the Musée des Arts et Métiers treats this eulogy as Vaucanson's official biography and reprints it in a pamphlet called "Jacques Vaucanson." According to Condorcet, Vaucanson leaves the provinces and arrives in Paris, recognizes that he is talented, gives himself entirely to his work, and hopes for well-deserved success.[62] Condorcet even conjures up a hostile and beknighted uncle, who is opposed to Vaucanson's pursuit of his métier. Condorcet does not fail to comment on the nature of the *grand monde* of Paris at the time. This was a world *avide de nouveauté*, or hungry for novelty. What Vaucanson will do is try to realize his ambitions by satisfying this hunger. Condorcet was able to recognize the ways in which Vaucanson played to and satisfied the appetite of the *grand monde* for novelty and entertainment. For Condorcet, Vaucanson's ability to attract the interest of a lay public did not in any way mitigate his scientific achievements.

The story of Vaucanson's arrival takes part in the formation of modern science in its relationship to *technique*. Science is privileged over technology and is thought to have preceded it, but as Heidegger reminds us, "Techne is the name not only for the activities and skills of the craftsman, but also for the arts of the mind and the fine arts. Techne belongs to bringing-forth, to poiesis; it is something poietic."[63] Samuel Weber emphasizes Heidegger's reversal of traditional notions of causality: "Science, [Heidegger] argues, depends both in its principle as in its practice upon *Technik*, rather than the other way round, as is generally thought."[64] In "Upsetting the Set Up: Remarks on Heidegger's Questing after Technics," Weber explains that the problem of the relationship between technology and art is articulated by Heidegger as one between *technē* and *poiesis*. This can also be understood as a relationship conditioned by a disunified temporal, non–goal-oriented field in which usefulness is not necessarily progressive and progress not necessarily useful. It is this distorting time that allows us to understand technology as never purely utilitarian: Heidegger reminds us that there is a primordial relationship and difference between the *technē* of technology and the *technē* of art. In mid-eighteenth-century France, one can imagine that the Classical relationship between *poeisis* and *art* to *technē* that Heidegger emphasizes in "The Question of Technology" is not yet fully obscured by the acceleration of technological progress. Heidegger's reading of *technē* and technology[65] radically revises the chronological and therefore sequential order of the history of science:

Chronologically speaking, modern physical science begins in the seventeenth century. In contrast, machine-power technology develops only in the second half of the eighteenth century. But modern technology, which for chronological reckoning is the later, is, from the point of view of the essence holding sway within it, the historically earlier.[66]

The automaton appears precisely in the time warp between the birth of modern physical science and the appearance of machine-power technology. The reversal of chronological order in the advent of modern science is only one of technology's special effects. It is a temporal distortion that Benjamin was ready to address, albeit in a different way, when he insisted on the relationship between historical materialism and the dotted time line, the disjunctive temporality of countdown in messianic time.

Madame de La Poupelinière / Thérèse des Hayes

The following is a story intended to supplement this brief biography of Vaucanson with an account of a series of events that occurred around the milieu of Le Riche de La Poupelinière. This young woman's story of success and failure might change our perspective on Vaucanson's arrival, insofar as it enriches our understanding of the problem of sexual difference in midcentury Paris. Ambition is what the talented young man and a gifted young woman share, but there is a vast difference in the realization of their respective destinies.

In Thérèse des Hayes's story, there is a wealthy libertine who is fond of opera girls, giving lavish parties and pursuing the pleasures of a wealthy bachelor. There is an attractive, intelligent young woman without any means, who turns out to be not so helpless after all, brilliant, beautiful, but too hubristic. There is a marriage, an adulterous affair, and a very unhappy ending. This story could be told as one that narrates the self-destruction of a resourceful, passionate woman who was not simply a victim like the hapless Justine, nor a cold-hearted villainess like the ruthless Juliette. What we know about this heroine we cull from a series of supposedly cold, hard facts.

La Poupelinière was a farmer general whose devotion to the arts of theater, dance, and music was only amplified by his passion for the countless actresses and girls of the opera. Grimm, Voltaire, Buffon, the Duke of Richelieu, the Maréchale de Saxe, as well as important government officials like Bertin were his regular guests. Most important of all, Rameau was a fixture in La Poupelinière's circle and presided over it in a position of great respect. Around 1734, the heroine of this tale, Thérèse des Hayes, was given by her mother, the actress Mimi Dancourt, to La Poupelinière to educate and take care of.[67] In return, the mother received a sum of money. Thérèse, however, was not content to remain the chattel of a rich man. In spite of this, the young woman somehow prospered in La Poupelinière's world as his concubine and mistress; she studied music with Rameau, and her interest in the sciences and philosophy earned her Voltaire's respect. Des Hayes remained a great ally of Rameau's and passionately took his side in his debate with Rousseau.

In 1737, Thérèse des Hayes found herself pregnant with La Poupelinière's child. La Poupelinière was not inclined to marry her, so des Hayes had her brother, the secretary of the cardinal de Tencin, appeal to Tencin's sister, Madame de Tencin to take up her cause. Madame de Tencin was able to persuade the Minister de Fleury to come to Thérèse's aid. The minister threatened La Poupelinière with the loss of the farms from which he derived his income and in this way forced the libertine to take des Hayes as his wife. In October 1737, the two were married. At this time, Vaucanson was already a regular guest in the La Poupelinière household.

We find, in a footnote to the Graffigny correspondence, the following account of Françoise-Catherine-Thérèse des Hayes: "For ten years, she presided over her husband's salons, at Passy and in Paris. Her relationship with the Duke of Richelieu led her husband to separate from her."[68] In her correspondence Graffigny describes her sense of isolation in her employment for the Duchess of Richelieu and gives voice to her desire for the duchess to introduce her to La Poupelinière and his wife. Clearly, from Graffigny's point of view, Thérèse presided over one of the most famous and animated of Parisian salons; she only enhanced her husband's already great reputation for entertaining and running a house and salon where the arts were honored and celebrated.[69]

While she presided over the salon of her husband, Madame de la Poupelinière was courted by "les plus fameux séducteurs de Paris."[70] Less cautious than the Princess of Clèves and more ambitious than Graffigny's Peruvian princess, Zilia, Thérèse seems to have accepted the Duke of Richelieu as her lover in 1744. Like the Prince of Clèves, La Poupelinière had his wife watched and upon receiving confirmation of his suspicions, he became physically violent with her, and she went to the police. The jealous husband then sequestered his wife, but the Duke of Richelieu was so enamoured of her that he found a way of renting the apartment adjoining hers. He had his workers devise an opening in the fireplace that was connected to

Madame de la Poupelinière's music room. The lovers were, by means of this secret door, able to continue to meet, but not for long. The husband's suspicions were not allayed, and one day, during her absence, he made an inspection of his wife's apartment, accompanied by friends, among them Vaucanson.[71] In Jean-François Marmontel's version of the discovery of the secret door, Vaucanson is given a buffoonish role. He admires the work of the hinges and springs of the secret door, while La Poupelinière grows more and more furious about the entire situation. The three men enter the rooms of Madame de la Poupelinière, and a member of the party notices that even though the weather was cold, there were no ashes in the fireplace. Vaucanson is the one who notices that the back of the chimney is mounted on hinges and is actually a very well-fitted door. In what is undoubtedly a much embellished version of the story, Vaucanson as an engineer is overcome with admiration of the chef-d'oeuvre that is the hidden door. La Poupelinière is indifferent to the door as mechanical marvel. For him the trick door is merely a sign that the sequestered wife has succeeded in outwitting him:

> "Oh Monsieur!" [Vaucanson] exclaimed . . . turning toward La Poupelinière. "What a beautiful piece of work I see there! An excellent craftsman must have made it! This plate is mobile, it opens, the hinge is of such a delicacy! . . . There is not a snuff box that is better made. . . ."
>
> "What, Monsieur?" says La Poupelinière, growing pale. "You're sure that this plate opens?"
>
> "Yes, I'm absolutely sure of it. I see it clearly," says Vaucanson, overwhelmed with admiration and pleasure. "I've never seen anything more marvelous."
>
> "What do I care about your marvel?"[72]

La Poupelinière, despite Vaucanson's protests, calls for workers to force open the door. In Marmontel's story, the comedic effect is produced by a radical disjunction between Vaucanson's and La Poupelinière's perspectives: the engineer is blind to the door's significance to a cuckolded husband, and the husband cannot see the door's craftsmanship. What this anecdote does is satirize the blindness of both men; it is a trick door that is only partially visible to each man. Of course, the reader joins the narrator in a conspiracy of insider knowledge: what the reader can see is that the door is both marvelous and an outrage.

Vaucanson expresses regret that such a singular masterpiece should be destroyed, but the jealous husband wants the contraption to give up its secret, and the only way that he can extract any certain knowledge from this machine of duplicity is through its destruction. The men in this story do not look to the woman for evidence of her desire and infidelity but, rather, to the space she inhabits: the cold, clean hearth betrays her. The absence of ashes in the fireplace is noticed and interpreted as significant: this significant lack requires a closer look. The closer look reveals a hinged door, a contraption, a machine that is evidence of illicit communication between the rooms of the wife and the rooms of another.[73]

For the Duke of Richelieu, this incident enhanced his reputation as a *galant homme;* for Madame de La Poupelinière, it proved quite fatal. The "machine" here is a contraption, a trick door: it provides for a way of foiling the interdictions of a jealous and brutal husband. For the duke it served to procure the pleasure of visiting his mistress; it did not serve to help the wife escape a brutal husband. La Poupelinière was afraid to confront the duke, and so he vented his anger on his wife, who without family and other means of support found herself in a position of dangerous marginality. She lived for a few more years, rather miserably, on a pension that her husband paid.[74] She eventually died, it was said, of complications somehow related to her husband's physical violence.

As in so many tales of libertinage, the duke's glory is paid for by the woman's destruction, but this story is not necessarily a completely dystopian one. Nancy K. Miller, in *The Heroine's Text,* demonstrates that there were two fates reserved for heroines of the eighteenth-century novel: marriage or death. Marriage, or what Miller more specifically calls "the heroine's integration into society," is the euphoric feminine destiny; "death in the flower of her youth" describes the dysphoric one.[75] Thérèse des Hayes's story is both "euphoric" and "dysphoric." In one sense, she is fabulously successful at securing her rights as a woman: she not only marries the libertine, she cheats on him. Her rapid ascent from her precarious status as a young woman "practically sold" to the triumphant adornment of her husband's salon and home is a reflection of her intelligence and remarkable resourcefulness.[76]

Having risen so far, however, she did not become more cautious and was in fact reckless twice over in her affair with the Duke of Richelieu, the gallant husband of Madame de Graffigny's unhappy employer. Thérèse des Hayes, it seems, was both a passionate and ambitious woman. If we extend the frame of her story to her disgrace and "miserable" death, she becomes another victim of the double standard of sexual comportment that doomed a passionate woman to either a sterile life of nun-like renunciation or a disordered life of ignominious self-indulgence. Des Hayes's fate is doubled and ambivalent, for she is both triumphant and defeated. The marriage that she is able to make so ingeniously with a wealthy libertine, almost her master as she is almost his slave, is finally the death of her. Femininity is like a hinge on which the truth of interpretation turns, hence its relationship to the ambivalence of knowledge itself. Des Hayes's story confounds the certainty of singular interpretations; her life falls into and exceeds the limits of literary representations of feminine destiny.

Like her husband's faithful Vaucanson, des Hayes was able to rise from rather obscure and dubious origins and succeed in penetrating the gilded salons of mid–eighteenth-century Paris. Des Hayes, however, wagered everything on a love affair; she died the death of a disgraced woman. Vaucanson continued to work hard, took few risks, married his daughter to an aristocrat, and died the death of a respected bourgeois, a full member of an increasingly powerful class. Des Hayes died the operatic death of a tragic heroine who has completely lost her position in the social order. Her fate hinges on a secret door, a contraption, a device, a small mechanical marvel. The secret of the rooms to which she has been confined is betrayed. The knowing eye of an expert and the anxious eye of a jealous husband are her undoing. The fireplace is, of course, at the hearth of the matter. Madame de la Poupelinière's fireplace is also remarkable for a lack—the absence of ashes during the winter months. The secret contraption, the false back, indicates illicit communication with other rooms and other bodies. That is, it is the sign of a secret whose discovery will have disastrous consequences for the woman involved.

Indeed, the fireplace has held significance for other critical readers, in particular Marie Bonaparte and Jacques Lacan, for whom it becomes the key to a psychoanalytic reading of Edgar Allen Poe's *The Purloined Letter*.[77] The content of the purloined letter is never revealed, but its circulation among many hands is the story that Poe's narrator tells. For Bonaparte and Lacan, the letter lying between the legs of the fireplace is the sign of castration, sexual difference, and, we shall add here, desire. The fireless fireplace, the place without fire, is the place of Madame de la Poupelinière's access to desire. As Freud has shown, in dream work the machine is often a stand-in for the genitals, whose susceptibility to manual manipulation makes it an easy metonymical displacement. The woman's place at home and hearth is a fireplace: in Thérèse's case, however, the fire is displaced. She tends other fires, set by her lover, in other places. Burning for each other, the duke and she pass through the false back of a fireplace. What for him becomes an amusing story of his gallantry and ardor becomes for her a trial by fireplace.

Getting Ahead with Machines?

The betrayal of the fireplace deprives Madame de la Poupelinière of her privileges in marriage and her position in society: it puts her back in her place—that is, out on the street, where she becomes again a daughter of a disreputable actress. Des Hayes makes good her first escape from this ignominious state by changing places and elevating herself through the influence of her friends to coerce La Poupelinière into a marriage. Marriage, however, is not enough for this gifted and charming young woman. She does not know the vulnerability of her place and will not behave like a woman who is entirely dependent on her husband or her marriage for her social standing and economic security.

Science and scientific interest guaranteed Vaucanson, *un moyen de parvenir*, the king's sympathy and eventually recognition from the Académie Royale des Sciences. The ambitious young woman, Thérèse des Hayes, must depend on a different and stranger form of science for feminine survival: her ability to set certain forces into motion that will allow her to make an advantageous marriage. It is a knowledge of human vulnerabilities and frailties that Walter Benjamin attributed to the courtier, the one who is the master of the intrigue, who plots behind the scenes and who knows how to exert influence indirectly, or from afar. Like the best of courtiers, Madame de la Poupelinière is finally doomed by her own taste for intrigue and confidence in machinations.

Although their means of arrival are so different, Vaucanson's and Madame de la Poupelinière's respective destinies and destinations are linked by a secret door. The door is a machine of deception, a machine that deceives everyone who uses it into believing that there is a singular significance for this device. This door is like the door to a crypt, a place where a secret must be kept and never let out. Machines displace notions of difference and representation when they occur in literature and history, especially when we find them, hidden away as ingeniously constructed trapdoors in the back of fireplaces in the music rooms of sequestered eighteenth-century wives. Thérèse des Hayes's story provides us with a new perspective on one of the figures that Derrida uses to designate *différance*, or *la brisure*—a break or joint that reveals itself to the suspicious eye as a sign of desire and treason. The hinge on which difference hangs facilitates the illusion of a homogeneous surface that can nevertheless, with the right amount of pressure, be broken. This break is an opening, an opening up and an opening toward, an understanding of difference as a difficult-to-represent device on which representation turns.

Chapter 5
Don Juan Breaks All
His Promises but Manages
to Keep One Appointment
(with History)

IN PRESSING ON, IN ORDER TO IDENTIFY the joint or break that modernity has made with the past, and in order to reassess a reading for the mark of violent historicity, we find ourselves considering a literary figure whose fall is the subject of great debate in the world of comparative literature. What follows is first and foremost based on a play on words that promises nothing more than the production of a few critical insights, and a modest amount of intellectual pleasure or irritation, depending on the position of the reader. The modesty of such a promise may seem highly calculated and perhaps should not be taken at face value.

I would simply like to propose that what is missing from the debate about Don Juan or DJ is JD or Jacques Derrida. Their initials are mirror images of each other, and they both seem to suffer similiar fates: vilification or adoration, with some subjects vacillating between the two. In Molière's play, *Dom Juan*, Don Juan plays with language, does things with words, seduces a few women, hypocritically declares himself converted, and avoids paying his debts. The figure of DJ is in our contemporary context, increasingly linked to questions of musical citations, appropriations, remixing, and the abuse of original material. Lighthearted and light-fingered, DJ produces nothing original or material except a good time. Don Juan gives a series of virtuosic and highly coded performances that are dependent on the disruptive potential of citationality itself: we could even call him a deconstructionist *avant la lettre*. On the other hand, Jacques Derrida could be said to be a Don Juan of philosophy whose promiscuous relationship with literature mixes everything up for everyone. His work has continued to infuriate some,[1] while seducing a few others. Deconstruction has had a less-

106

than-reverential attitude toward traditional practices of philosophy and literary criticism. If it demands that we pay a different kind of attention to reading and writing, it is because deconstruction has reassessed its outstanding debts to traditional philosophical and critical filiations.

Don Juan refuses to honor both his debts and his promises of fidelity. When the infidelity of the consummate seducer is dramatized and theatricalized, a showdown between literature and history takes place in the background and sets the scene for provisional victories and long-term concessions. Don Juan plays around by refusing to be pinned down, but the dramatic tension is sustained because the stakes of his playing around continue to get higher as the play unfolds. In the theater, knowing how to keep one's place is crucial to dramatic continuity. Don Juan's displacements produce a theater of radically disrupted places where promises are made but not always kept.

The disturbances of place in the play are very much contingent on the fact that as a dramatic character Don Juan inhabits a temporality of amnesiac precipitation: he breaks his promises, but above all, he breaks off all relationship of indebtedness—not only with his past but with *the* past. He lives fast and remains at odds with any logic of temporal progression or continuity: in so doing, he is out of time and he chooses a spatiotemporal exclusion. Molière's Don Juan offers himself up to be read as a force resistant to both history and historicization; this force is inherent in and deployed by the disjunctive temporality of the literary object. In Molière's play, anachronism as a literary or rhetorical figure dramatizes power of radical untimeliness and violent decontextualization. Oscar Mandel expresses one popular critical view of this figure when he writes, "One could write a useful literary history of Western Europe by citing for illustrations nothing but Don Juan texts. Don Juan never created a climate; he always responded to it, and in fact responded to it with almost academic perfection."[2] Mandel implies the commonsense idea that literature can be so "in tune with history" that a literary figure can actually illustrate the unfolding of literary history itself. What literature illustrates is a historicizing logic that is founded on evolutionary specularity. The evolution of literary works can be mapped according to a homogenizing temporality of periodization. If Mandel suggests that Don Juan is the perfect academic, he implies at the same time that academic work should be "a perfect response" to historical "climate." Academic literary criticism, then, is no more than historically determined weather reporting. For Mandel, Don Juan is a very good weatherman, not predicting the weather of the near future but illustrating the "climates" of the distant, and not-so-distant, past. More than just a good historical reporter on climatic changes, he is a reflection of the progression of literary history itself.

It is hard to imagine that Don Juan can be rehabilitated as any kind of academic, but it is seductive to us to see him as an academic. For Georges Poulet, Molière's theater dramatizes the comic break in temporal and moral continuities in order to seal all the more securely the sense of an "eternally valid judgment on human deportment." The spectator, for Poulet, is all the more tightly bound to the continuous order of things when he or she witnesses the dramatic exclusion of the comic character. For Poulet, there are two kinds of "duration": "The duration of the order in which one participates and lasts; and the instant of disorder which is limited to the object and which interrupts time."[3] Poulet is optimistic about the containment of disruptive interruptions and gives "eternally valid judgment on human deportment" the very last laugh. The attempt to make an eternally valid judgment interrupts Don Juan's interruptions: the critic would like to stop the libertine cold in his tracks and cut him out of the duration of criticism altogether. The encounter between the critic and Don Juan can be seen as allegorizing a confrontation between critical insight and the force of literature. Mandel and Poulet promise to answer the question "Where does Don Juan belong?" For Mandel, he is firmly grounded in his historical period, as it has been defined by the discipline of comparative literature; in the case of Poulet, he is the absolute outsider of duration, continuity, and community. For Mandel, Don Juan belongs to us as historicizing academics; for Poulet, he is that which is most radically excluded from the community of critics, readers, and spectators. These two critics can reach diametrically opposed conclusions about the nature of this literary figure and remain in absolute agreement with regard to the place and promise of literary criticism. For both, the critic draws the spatiotemporal grid on which the literary figure will be placed and understood.

For Benjamin, the force of criticism is located on the ground of its discontinuity with the temporality of perception and analysis: the belatedness of literary criticism with regard to the literary work forces on it totalizing strategies of mimesis (plot summary with commentary) or fragmentation (close reading).[4] De Man's criticism offers a disturbing account of temporal continuities as criticism's preciously guarded illusion. Armed with such an insight into the deceptive nature of critical time, a critic cannot escape the dilemma of her own parasitic activities, but such an awareness does offer an exhilarating freedom, described by Carol Jacobs in much the same terms one could use to describe Don Juan's own character:

> Its [criticism's] time is... an act of transgressive freedom, a rupture that marks the impossibility of textual definition and self-definition. It performs this deception with respect to the texts it reads and also with respect to the text it cannot and yet inevitably does read, itself. It acts out, then, both the promise of progress and its failure, making promises it cannot fulfill in the present, making excuses rather than confessions for that which it might rather expose than hide, narrating endless fictions.[5]

In making promises he cannot fulfill, Don Juan himself demands not so much identification/empathy or judgment as interpretation/analysis. He asks to be read, and with a certain amount of coldness. Benjamin identifies the nostalgia of historicism as a melancholy that tries to suppress the violence of the present in order to empathize with origins; Don Juan defies this kind of "indolence of the heart."[6] The scene of Don Juan's immolation is supposed to provide for a spectacle of punishment and exclusion, but perhaps the dramatic pathos of Don Juan's fall is too intense: "Oh Heavens! What do I feel? An invisible fire burns me. I can't stand it any longer. My whole body has become an ardent fire."[7] If Don Juan is finally consumed by the flames of a passion to which he has previously proven to be immune, his exclamation of suffering, expressed in the highest tones of melodrama, seems designed to inspire a kind of Aristotelian pity and fear that is more proper to the demise of the tragic hero or martyr. It is a belated appeal to pathos that cannot sustain itself. Sganarelle's exclamation, "My wages! My wages?" provides a comic footnote to Don Juan's demise. If Don Juan is burning, his fate produces a certain coldness and detachment. It is Sganarelle's low venality that provides for the dramatic cut that protects an audience from infelicitous identifications.

If there is something unresolved about Don Juan's punishment, it is because he remains an enigmatic figure who cannot be made to stay in one place. This enigmatic quality seems eminently literary. Literature somehow manages to wage a strong resistance against the force of contextualization, even if it does not succeed in resisting it entirely. In Molière's play, anachronisms are used to represent the violence of radical untimeliness and decontextualization. Don Juan's duplicity has a lot to do with his ability to arrange spatiotemporal aberrations. Anachronism as a literary device is acceptable, but anachronism in criticism is always characterized as error.

The error of anachronism is based on a transgression of the boundaries of periodization, but this historicizing tendency is itself embedded in a complicated history of its own. In the historicist view, the present always runs a deficit with regard to the past, and intellectual work is about keeping up with interest payments. An anachronistic reading has fallen short on making late payments on the outstanding balance in literature's account of history and context.

Criticism that threatens the coupledom of literature and history is usually characterized as formal and apolitical. The notion of political engagement has been inexorably related to the idea of a responsibility to history. In order to add some color to and fill in some details of the corridor wars being fought within departments of literature, we can make a rough sketch of the situation in the following way. The idea behind historicizing literature is based on the fact that literature and history have always been compatible; history is the more stolid member of this interdisciplinary couple. This union is seen as progressive: for too long, literature has been held back from history. Generic divisions (poetry, drama, prose) effaced historical differences in the name of formal categories, but now historical divisions (usually named after centuries) tend to have gotten the upper hand as literary studies have been gradually reconfigured according to historicizing narratives. Literary criticism is a house that continues to be divided against itself. Within what can be characterized as the formalist division of literary studies, a crucial subdivision exists: there are the theorists (often influenced by deconstruction) and the unreconstructed formalists. The theorists are accused by both historicists and formalists of being generally irresponsible to scholarship itself. While formalists and historicists accuse each other of crimes of infidelity—either infidelity to historical context or infidelity to the text itself—the theorists are accused of having betrayed the contract of a humanist education *tout court*. What does fidelity mean or imply when it comes to reading, writing, and other scenes of mediated communication?

The following passage from an article by Larry Riggs both sets out the problem and then tries to offer solutions to it. The interpretation of Don Juan is what is at stake:

> The twentieth-century tendency to see in Dom Juan a hero of anti-conventional "authenticity" is anachronistic: in an elaborate, hierarchical social structure such as that of seventeenth-century France, the self is not separable from its roles and accoutrements. Honor, nobility, and

ethical veracity depend on deserving to be oneself—on meriting one's "costume."[8]

According to such an argument, twentieth-century critics who are eager to indulge in the tendency to see in Don Juan a hero of authenticity (in Mandel's case, this authenticity might take the form of being a good academic) are in fact tearing him from his context, a context in which the "self" is configured entirely differently from the manner in which it is constructed in the twentieth century. Such tendencies toward anachronism can be tempered with a good dose of historical understanding. History will permit us to engage in a consistent recontextualization of literary object, which it seems is all too easily torn from its "times," its setting. Riggs's notion of a "seventeenth-century" self, entangled in a web of social relations and social obligations, is certainly not inaccurate, but could we not continue the round-robin of countercorrections by asserting that the term *seventeenth-century self* is yet another twentieth-century abuse of a seventeenth-century problem? Accusations of anachronism can happily proliferate as one critic can always claim greater fidelity to history than his or her colleagues.

It is quite evident that for Larry Riggs his reading is superior to other readings of the play because he is more *faithful* to the historical context of Molière's play. If we honor history as context for literature, we are in a sense implying that literary criticism is fundamentally indebted to history. The ability to remain faithful or loyal is maintained through a sense of indebtedness. It just so happens that in this piece of literature, however, we find an attempt to cancel all outstanding debts. As a son, Don Juan demonstrates no indebtedness toward his father. As a noble, he cultivates no loyalty toward his distinguished genealogy; as a lapsed Christian, he is not terrorized into any sense of obligation toward God; as a lover, he feels no gratitude toward women; as a debtor, he feels no obligation to pay his creditors. This particular brand of unaccountability and faithlessness can be understood simultaneously as violently reactionary[9] and terrifyingly progressive in his seigniorial disdain for contractual arrangements: "It is very bad politics to hide oneself from one's creditors. It's good to pay them something, and I possess the secret of being able to send them away satisfied without giving them a cent."[10] He is exercising an increasingly ineffective feudal prerogative, but in his lucid and lighthearted atheism, he is happy to practice a radically logical, secular, libertine reason: "I believe that two and two are four, Sganarelle, and that four and four are eight."[11]

Molière's Don Juan has been taken up as an exemplary literary figure by Shoshana Felman, Sarah Kofman, and Jean-Yves Masson, but in a way that breaks with Mandel's and Poulet's work insofar as there is no evidence of the taking of historical temperatures or the passing of judgment. At first glance, the absence of aesthetic appreciation and moral condemnation might seem insignificant, but it marks an important rupture in the development of literary theory itself. Kofman, Masson, and Felman show a certain chilly detachment in their refusal to read Molière's play as a lesson in either history or morality. Kofman and Masson show that Don Juan tries to "buy" more time from the ultimate creditor, God himself, in order to defer the settling of *all* accounts.[12] His refusal of all indebtedness is based on a deferral and a postponement of the payback that he is going to eventually receive. For Shoshana Felman, Don Juan is only playing with the effects of *pure* performativity: his promises of marriage are misfires in the Austinian sense because there is something missing from the context/conditions of such utterances. According to Felman, it is significant that J. L. Austin (in a sense, like Don Juan) uses marriage vows as an example of how *saying* something is *doing* something.[13]

Austin is extremely sensitive to the myriad of ways in which the speech act can go very, very wrong. Marriage, like all performatives, is vulnerable to many infelicities:

> Speaking generally, it is always necessary that the circumstances in which the words are uttered should be in some way, or ways, appropriate.... Thus for naming the ship, it is essential that I should be the person appointed to name her, for (Christian) marrying, it is essential that I should not be already married with a wife, living, sane and undivorced [Elvire] and so on.... Surely the words must be spoken "seriously" and so as to be taken "seriously"?... I must not be joking, for example, nor writing a poem.[14]

The problem with Don Juan, however, is that he might just be joking and could very well be "writing a poem" out loud, or at least citing one—within the context of a play, to boot. The question of his sincerity is certainly a difficult one, and one that he delights in rendering more difficult. Don Juan, in fact, basks in the enjoyment procured by promises that are *not* made seriously. Don Juan may be a bigamist and a liar, but he knows how to make a misfire work to his advantage.

In the strictest Austinian sense, a lack of seriousness (making a joke) and the making of literature (writing a poem) are aberrations in the use of "everyday" language. Making a joke of marriage, however, seems to be one of Don Juan's specialties (Sganarelle to Gusman: "You tell me that he has married your mistress: believe me that he would have done a lot more for his passion and that with her, he would have married you, her dog, and her cat.")[15] A marriage costs him nothing to contract. Every day, however, there are people who may or may not be married already who say "I love you" or even "I marry you" to a third party in the absence of a feeling of love and with no intention of following through on the contract of marriage. Yet we would not say that these actors or actresses are either liars or bigamists. They are in the theater. Linguistic performance is always susceptible to theatricalization. That Don Juan accepts his entire field of action as purely theatrical makes of him a particularly interesting literary case study. In a sense, he allegorizes the repetitive principle of theater itself. For Max Vernet, reading Molière is not a simple affair, precisely because a play is not like any other piece of writing: the specificity of the theatrical text calls for a kind of attention that accounts for the instability of the written trace as a repository of both accident and repetition.[16]

To return to the question of speech acts, the problem is that any speaker at any time can be simply *acting*. What acting might in fact be is, of course, highly problematic in all cases, but when we are dealing with actors we are perhaps adding to the confusion of agency: an actor follows a script and a nonactor supposedly does not. But it is certainly possible to imagine that a certain amount of acting takes place far from the stages of the world. In any case, Felman works through Molière's play in a discussion of Austin's speech act theory because, she reminds us, Austin was, like his fictional counterpart, less preoccupied with truth and falsehood than with felicities and infelicities; it is possible to derive from Austin's work an account of linguistic functions that are completely free from any responsibility for reporting on an empirical state of affairs.

There is one small problem. Austin, as Derrida points out, *excludes* the case of literature from his analysis of the performative. I cite Austin again:

> As utterances our performatives are also heir to certain other kinds of ill which infect all utterances. And these likewise, though again they might be brought into a more general account, we are deliberately at present excluding. I mean, for example, the following: a performative utterance will, for example, be *in a peculiar way* hollow or void if said by an actor on the stage, or if introduced in a poem, or spoken in soliloquy. This applies in a similar manner to any and every utterance—a sea-change in special circumstances. Language in such circumstances is

in special ways—intelligibly—used not seriously, but in ways *parasitic* upon its normal use—ways which fall under the doctrine of the *etiolations* of language. All of this we are excluding from consideration. Our performative utterances, felicitous or not, are to be understood as issued in ordinary circumstances.[17]

It is, then, strictly speaking, impossible to find either a felicitous or infelicitous performative in the realm of literature and theater because Austin restricts the field of his analysis to "ordinary circumstances." The literary performative is a parasite that saps the ordinary performative of its original power; it drains its strength. This is not in any way an attempt to contradict Felman's contention that Molière's play is in fact constructed as a series of literary speech acts. These speech acts are hollow and void "in a peculiar way"; being framed in a theatrical work, they can be considered representations or, in Austin's words, *parasites* of the hollowness and emptiness of the ordinary misfire. But they should not be read as simple misfires in and of themselves: they dramatize or represent theatrically, the effects of parasitism itself.[18] Don Juan is perhaps no more and no less than a theatrical parasite who sets up peculiar performances of hollow promises.

Returning to Felman:

> What is really at stake in the play—the real conflict—is, in fact, the opposition between two views of language, one that is cognitive, or constative, and another that is performative. According to the cognitive view, which characterizes Don Juan's antagonists and victims, language is an instrument for transmitting *truth*, that is, an instrument of knowledge, a means of *knowing* reality. . . . Don Juan does not share such a view of language. Saying, for him, is in no case tantamount to knowing, but rather to doing: acting on the interlocutor, modifying the situation and the interplay of forces within it. Language of Don Juan is performative, not informative.[19]

Felman draws up a provocative list of Don Juan's duplicitous performatives: he says to Don Carlos, Elvire's brother whose life he has just saved, and who does not recognize him as his sister's abuser, "I am a friend of Don Juan . . . and I promise that I will help you get satisfaction from him. I commit myself to have him brought to the place that you will designate. . . . I answer for him as for myself."[20] To his creditor, M. Dimanche: "I am your servant, and furthermore, your debtor. . . . I beg you again to be persuaded of the fact that I am entirely at your disposition, and that there is nothing in the world that I would not do in your service."[21] And with the women he seduces: "I take as witness the man whose word I give to you.... I reiterate to you the promise that I have made.... Do you want me to make terrifying oaths? May Heaven...."[22] Certainly, Don Juan is guilty of a lack of seriousness and his speech acts are hollow in a peculiar way: what Molière's play dramatizes are the consequences of this kind of playing around. Everyone, from Don Carlos to M. Dimanche to Mathurine and Charlotte, is taken in by Don Juan's hollow promises and infelicitous speech acts. The audience is a delighted and occasionally outraged witness to the efficacy of such playing around (with language). In fact, Don Juan's infernal machine works so well that it takes no less than a deux ex machina, in the form of the Commander's statue, to settle accounts and set things right at the end of the play.[23]

When Derrida reads Austin, he emphasizes the fact that the excluded literary moment that has been judged to be parasitic is actually structurally necessary for any kind of pure performativity (under ordinary circumstances) to function successfully. Because of the constraints of time and space, I will resume some of Austin's schemata and Derrida's arguments in what will no doubt be a crude and simplifying way. For Austin, the speech act is felicitous when a certain set of conditions is fulfilled; these conditions can be divided into roughly two sorts. The first are composed of conditions of conventionality, and the second conditions of intentionality: "There must exist an accepted conventional procedure having a certain conventional effect, that procedure to include the uttering of certain words by certain persons in certain circumstances."[24] According to Austin, then, we must follow what convention dictates in order to successfully perform a speech act, for example, the naming of a ship or the taking of a wife (or, we might hasten to add, a husband). Thus the success of the conventional speech act is dependent on its recognizability as a *repetition* of other similar acts. This recognizability is encoded as a function of the context of a speech act. Derrida asks, "Could a performative statement succeed if its formulation did not repeat 'coded' or iterable statements, in other words, if the expressions I use to open a meeting, launch a ship or a marriage were not identifiable as conforming to an iterable model, and therefore it they were not identifiable in some way as 'citation'"?[25]

Derrida does not fail to imply that iterability is one of the conditions of literature. Literature operates on a principle of radical citationality: and the play, or the theatrical representation, is in fact even more susceptible to this principle. In French, to rehearse is to repeat *(répéter)*, and this certainly gives way to the possibility of a rich play on words, but it also describes one of the conditions of theater itself. The literary performative in particular is susceptible to citation, and this creates disturbances in the field of performativity in general. Theatrical representation is dependent on aberrations of repetition.[26] Most of the conflict between John Searle and Derrida takes place over the notion of what Searle calls "pretended" speech acts. Searle criticizes Derrida for making too much of Austin's "strategical" exclusion of promises made by actors on a stage or characters in a book because they are not "standard cases." Searle suggests that Derrida wants to begin a discussion from the place of the nonstandard, pretended, theatrical, literary, and parasitical promise, one that might be found in the text of Molière's play, for example. If Searle understands Austin's exclusion of the literary, theatrical speech act as merely a deferral, he does not really offer evidence of the final encounter. If it is a strategic and temporary exclusion, it seems to be one that hopes for an infinite postponement of such a confrontation. Derrida responds to Searle by pointing out in "Signature, Event, Context," "beginning with theatrical or literary fiction, I do believe that one neither can nor should begin by excluding the possibility of these eventualities."[27] No one wants to begin with an analysis of the pretended, theatrical, or literary speech act, but its deferral has very different consequences for Searle and Derrida. In *Dom Juan,* the full force of what it means to pretend to promise is never realized all at once. In an absolute present of promising, there is no way to tell if it is pretended or not. It is only in the unfolding of the drama that the full force of DJ's deceptions becomes clear. Derrida argues for a condensation of the time line of deferral in order to take into account that the standard speech act is always paying interest to its nonstandard, doubly pretending who might be profiting, like Don Juan, from their association with "real-life" promises.

Don Juan is hyperliterary, but more than that he is hypertheatrical: by rehearsing performativity within the theater, he is a monster of citationality and parasitism. All his utterances take place in a sense between quotation marks. His every utterance is marked by citation and repetition. And if he, as Kofman and Masson insist, refuses all sense of indebtedness, he is not very good at giving credit where credit is due. He is very bad at citing his sources; he acts as if this were original material.

That one can very well say something and not mean it threatens the smooth functioning of speech acts. In such cases, saying something is certainly doing something, but doing something rather dastardly, because the second set of Austinian conditions that have to do with the presence of intentionality, sincerity, and seriousness is not fulfilled. According to JD, one of the qualities of writing has to do with its potential to make a radical break with the context of its own production. A broken promise, then, points ironically toward writing. Don Juan's final gesture of hypocrisy takes place as a series of quasi soliloquies on the subject of his conversion, during which he patches together a string of formulaic pieties, much to the horror of Sganarelle. Sganarelle's own objections to his master's libertine thought, however, are formulated in parodic distortions of the clichés of moral outrage.

In act 5, scene 1, when Don Louis comes to greet his son upon the news of his false conversion, Don Juan's speech, "Oui vous me voyez revenu de toute mes erreurs" (Yes, you see that I have been cured of my errors), is introduced by the stage direction "faisant l'hypocrite" (playing the hypocrite). This is what Christopher Braider has identified as a trait particular to seventeenth-century theater itself: the staging of hypocrisy can be understood as a commentary on the theatrics of representation itself, especially when this representation has to do with the highly charged question of faith.[28] But if the hypocrite's only intention is to deceive, he does so only by citing a highly codified rhetoric of piety. In order to recognize a hypocrite, one must be able to recognize that the hypocrite is staging a spectacle of sincerity with borrowed phrases: in the representation of hypocrisy, the audience, as Jacques Guicharnaud points out, is always double and doubled. The hypocrite as actor evokes two radically different kinds of interlocutors: first, there is an audience that "sees" through the act; the second is one that cannot. In *Dom Juan*, Don Louis occupies the place of this second audience. What the first audience sees is the blindness of the second, even if this second audience, as in the case of *Tartuffe*, is an audience of one, Orgon.[29] If what is staged in *Tartuffe* is Orgon's blindness, there is in *Dom Juan* a multiplication of blind spots, places from which the hypocrite is invisible: blindness is a condition of being seduced by the hero. If the hypocrite were entirely successful at duping everyone, there would be no interest in such a representation: it is because he fails to fool everyone all the time that his case becomes an interesting one. In addition, his successes and failures have consequences for all the problems surrounding the staging of sincerity.

The question of intentionality or seriousness is highly problematized in all cases of literary production: we have to point to the importance of the break that Felman's reading makes with previous psychologizing interpretations of *Dom Juan* and the literary object. So much of this criticism, an exhaustive analysis of which would be impossible and inappropriate in this context, is based on the opposition between *être* and *paraître*. The problem of representing sincerity is eminently literary: the literary object is something that is always already threatening the relationship between meaning and appearances, authenticity, and intentionality. This focus on authenticity is obvious in the criticism surrounding *The Princess of Clèves*. When Lionel Gossman's analyzes Molière's *Dom Juan* in *Men and Masks,* he focuses on a critique of the difference between the protagonist's "real being" and its utter contradiction "with the image he gives of himself." Unlike the critics who Larry Riggs criticizes for touting DJ as a hero of (in)authenticity, Gossman sees Don Juan as nothing less than a failed human being. One could call Gossman's practice anachronistic or ahistorical: his writing certainly does not take into account the highly ceremonialized social life of seventeenth-century France. Gossman's critique, however, is not out of place in another kind of time line: his text is saturated with the vocabulary of contemporary American psychotherapy. He wants to identify DJ as lacking emotional maturity: he is diagnosing a literary character with an intense passion, and in doing so he is treating the question of historical context a bit cavalierly, as indeed, DJ might do himself if he were making an appeal to a certain audience. Just like Gossman, DJ does refuse a certain temporal logic when it suits him: the critic wants to show where the literary character is deficient, thereby allowing for the possibility that literary critics can subsequently counsel literary characters on the path to self-improvement.

According to Gossman, "The imposture of Dom Juan is, however, no mere surface phenomenon. It reaches deep into his innermost being, into areas where he himself is no longer aware of it."[30] Don Juan's very *being* is judged as having been corroded by his posturing and his *im*posture. Don Juan, says Gossman, is failing at no less than being human: "Dom Juan can find no peace or happiness in a real relation with another human being"(47). DJ is not capable of "real" relationships with others: he needs help. The first step taken on the road to recovery might be the critics' diagnosis. We can imagine a Don Juan, having assimilated the vocabulary of existential psychology, attending sex addicts' anonymous groups, trying to have real relationships, and professing faith in psychic healing while he continues to avoid "happiness in a real relation with another human being" by breaking hearts and promises.

Felman writes about the broken promises of Don Juan as a series of speech acts that are felicitous (in Austinian terms) on the level of seduction and in the production of pleasure. His interlocutors are interested in knowing about the truth of Don Juan's words, but Don Juan himself is not interested in knowledge or communication. He is interested in seduction and persuasion: language is not an epistemological field for DJ—it is a conflictual and eroticized space of performativity. Felman reads for linguistic effects, Gossman for ontological affect. If Felman understands Don Juan as always already a theatricalized character, a character that for Jean Rousset embodies or incarnates Baroque "Inconstance," Gossman assumes that there is an existential failure on the part of Don Juan that makes him morally reprehensible. The vertiginous gap that separates the ways in which Felman and Gossman read Molière's *Dom Juan* can only be accounted for by what occurred between 1963 and 1980: the dissemination of critical theory that called for a questioning of the ways in which we handle the literary object and the fictional subject. Gossman takes authenticity for granted as the ground from which he can launch his judgments, just as Riggs understood history as a stable entity. If Felman is able to appreciate Don Juan as consistent in his logic, she is taking into account a certain Nietzschean attitude toward the criminal: "But a criminal who with a certain sombre seriousness cleaves to his fate and does not slander his deed after it is done has more *health of soul.*"[31] Larry Riggs and Lionel Gossman, despite their differences, turn out to have more in common than one would expect. Gossman's apparently anachronistic take on DJ's problems and Riggs's recourse to historical context converge around the question of value, identity, and meaning:

> Dom Juan dramatizes the refusal of ethical risk in such a way as to show that society can either be a stagnant, ceremonial game, circling toward total disillusionment, or a collective effort to conceive and realize values. As Dom Juan's deception reduces others' respect for him to the level of superstition, he became a phantom and analogous to the *loup-garou* of Sganarelle's fantasies. Ethical meaning and social value can exist in a context of substantial exchanges involving both risk and mutual benefit, and individual identity can be expressed and preserved only in ethically meaningful relationships.[32]

Riggs appeals to a limited sense of historical synchronicity: there are obviously certain categories that transcend temporal differences like "ethically meaningful relationships." His program for social harmony sounds engineered for mutual coexistence under late capitalism: how best to "conceive and realize value," that is, how best to maximize profit for the majority? The phrase *mutually beneficial social relations* partakes of the rhetoric of industrial bureaucratic societies where exploitation and violence are controlled and rationalized.

Riggs's appeal to ceremony suddenly falls apart when he writes that ceremony becomes *empty* games of "society," thus implying that there are meaningful and, more important, profitable social conventions and rituals that it would behoove us to respect. In *The Court Society,* Norbert Elias demonstrates that the so-called emptiness of court society only appears as such because our vision is occulted by bourgeois notions of "expression" and "individuality."[33] Elias tries to show how French court society under Louis XIV differed radically from the bourgeois social space. Elias convincingly demonstrates that the ceremonial games were the only games in town. What Elias calls "the struggle for prestige" played itself out in social spaces: the struggle for power took place exclusively along the lines of etiquette, ceremony, and convention. If we are in a sense prone to condemn the ethical and moral failures of court life, it is only because these ancien régime social formations continue to haunt us and must be repudiated and repressed as a recent past, whose tacky styles have to be hidden in the back of our collective closets. Riggs's careless use of the terms *individual expression* and *ethically meaningful relationships* is just the kind of bourgeois notion that is used in order to cover up the difference between court and industrial/bureaucratic formations. The bourgeois categories triumphed, especially after 1789, but in winning the class struggle, the bourgeoisie has not had any rest. It has ceaselessly rewritten history as the story of winners who are simply paving the way for the better world of "individual expression," "social cooperation," and mutually beneficial relations. Riggs's appeal to forming more "ethically meaningful relationships" smacks of a smug moralizing of the winners, which does everything it can to prevent a historical understanding of the price of its own victory. Winning becomes a kind of morality itself, and in such a world DJ is not only a bad egg, he is a loser. It is crucial for Riggs, as it is for Gossman, that the literary and historical aspects of Don Juan's nonphenomenological existence be mystified by judgments of value that promulgate a highly suspect set of ideals.

Don Juan's refusal of debts is a very specific kind of unethical behavior that attacks from two positions the very formation of morality: he is the arrogant aristocrat whose recourse to his position and pure force abrogates him from all contractual agreements. At the same time, he is the libertine free thinker whose disbelief in God is transposed onto his relationship with all antecedents and creditors. Nietzsche traces the origins of conscience and morality to the very problem of indebtedness, and he struggles to elucidate the complicitous relationship of *Schuld* (guilt) with *Schulden* (to be indebted). In so doing, he manages to cast doubt and uncertainty on the very constitution of meaningfulness and morality. Austin, DJ, and Nietzsche are not so interested in meaningful relationships: they are more interested in the force in language, characterized by Austin as illocutionary and/or perlocutionary, which creates differential relationships and felicitous or infelicitous effects.

We might be able to understand Gossman's and Riggs's moralizing as a refusal of indebtedness to Nietzsche.[34] In refusing Nietzsche's legacy in reading DJ, they can only find him a failure and a liar. The force of unreflected genealogy of moralizing in and outside of literary criticism demands to be read as a symptom of Nietzschean denial.[35] The moralizing tendency in literary criticism persists in a tenacious way, and it struggles to maintain the terms of its own confusion by placing unquestioned value on certain categories: truth, being, expression, meaning, individuality, substantial relationships, and so forth. Nietzsche's intervention can be deferred temporarily, but never completely.

Nietzsche recurs as a figure in Vernet's, Felman's, and Kofman and Masson's work on the play. Felman cites the Second Essay of *The Genealogy of Morals* in order to evoke the image of man as the promising animal: "To breed an animal with the right to make promises—is not this the paradoxical problem nature has set itself with regard to man? and is it not man's true problem?"[36] Kofman and Masson go further and demonstrates that Don Juan is no less than the embodiment of animal forgetfulness.[37] Citing Nietzsche: "Now this naturally forgetful animal, for whom oblivion represents a power, a form of strong health, has created for itself an opposite power, that of remembering, by whose aid, in certain cases, oblivion may be suspended, especially in the case of promises."[38] For Nietzsche, remembering is not punctuated by forgetting—forgetting is punctuated by remembering. Don Juan's willful forgetfulness with regards to all contractual relationships is a side effect of his healthy, animal-like indifference to the past, to debts, to promises:

> As the sovereign man, he disposes of time: he is not interested in the past (it is Sganarelle who is the accountant of Don Juan's conquests; he serves as his memory aid); he anticipates future conquests, makes plans for kidnapping a peasant girl who has resisted him, even if he is ready, very quickly, to abandon conquest for another, and is devoted to the

time of successive encounters, to the wonder of meetings, therefore to a discontinuous time in which one moment is neither linked nor responds to another.[39]

Forgetting has something to do with the nature of literature itself: we must remember that forgetting has a salutary effect on each renewed encounter with the literary text itself.

Sganarelle in Molière's play and Leporello in Mozart's opera are both like accountants, trying to keep count of their masters' conquests by recording them, writing them down. Don Juan (Don Giovanni) lives a disjunctive temporality, moving from chance encounter to chance encounter, from eternal present to eternal present. He is a son who will not acknowledge the privilege of ancestors or antecedents. He is the present realized and reversed at every moment. Don Juan is haunted by a Nietzschean relationship to history: the flames that consume Don Juan at the end of the play are reminders of his mortality, but Nietzsche gives DJ a second life of significant forgetfulness. DJ, it is becoming clear, is an allegory of the temporality of modernity, which strives to realize itself in the present tense of literature. There is something about this present tense that resists the kind of historical determinism that conceives of temporality as evolutionary, unbroken progress that can be retold as a homogeneous narrative of various climactic shifts and developments. The past is impossible to wholly reconstitute and therefore impossible to completely domesticate through the techniques of good bookkeeping, and the search for an empirical ground.

Ever since Otto Rank's study of the Don Juan legend, psychoanalytic theory has been seduced by the case of the aristocratic philanderer.[40] The psychoanalytic account sees in Don Juan the embodiment of a symptom. Monique Schneider describes him as being led back to a past from which he is attempting to flee. He is very simply described as a figure of the symptom of repetition compulsion. A purely psychoanalytic reading of the problem of DJ, however, remains as reductive and inadequate as a more general psychological one: it is certainly possible to read the temporal aberrations of the narrative as pathological, but we should not forget that what is framed by the play is most importantly the spatialization of a literary and theatrical temporality. The literary representation is a break in time that leads to a repetition of the time of the action of the play. Theatrical representation sets up the frame for the repetition of repetition.

DJ is he who does not belong: he is not of his time, nor of ours. He is the break with time and history. Kierkegaard describes him as the tireless, inhuman force of seduction: "He needs no preparation, no design, no time; for he is always ready. Indeed power is always in him, and desire too, and only when he desires is he really in his element."[41] For Kierkegaard, Don Juan represents something in literature that defies all forms of identification: as such, he is more than not human—he resists humanizing impulses. The desire for and of forgetfulness is also a state of constant preparedness for the present: force as desire to break with the past is given a name by Paul de Man in his reading of Nietzsche's *On the Advantage and Disadvantage of History for Life:* de Man calls it modernity. Nietzsche's ruthless forgetting, the blindness with which he throws himself into an action lightened of all previous experience, captures the authentic spirit of modernity:

> It is the tone of Rimbaud when he declares that he has no antecedents whatever in the history of France. . . . it is the tone of Antonin Artaud when he asserts that "written poetry has value for one single moment and should then be destroyed. Let the dead poets make room for the living. . . . the time for masterpieces is past." Modernity exists in the form of a desire to wipe out whatever came earlier, in the hope of reaching at last a point that could be called a true present.[42]

Modernity is Don Juan's other name: he is a figure who sets into motion a temporality of usurpation and impatience. It is astonishing how much Artaud's statement from *The Theater and Its Double,* cited by de Man above, echoes Don Juan's brutal ejaculation after his father's unexpected visit: "Ah! Die as soon as you can, it's the best thing you could do. Everyone has to have his turn, and it infuriates me to see fathers who live as long as their sons!"[43] DJ wants his time to be his own and no one else's. His wish for his father's death makes him all too likely a candidate for a convenient Freudian reading, whose very obviousness should make us suspicious. I think that it is more interesting to consider the crisis in filial relations that is represented in seventeenth-century French drama and compare him with his literary counterpart, another virtual patricide, Corneille's Don Rodrigue, who is torn between a sense of double indebtedness—to his father and his mistress. (Corneille's Rodrigue is El Cid, a man who is in love with a woman whose father has offended his own. In order to be loyal to his father, he must betray his lover and vice versa.) Kofman and Masson describe Rodrigue:

> He is the one who knows how to count and the one on whom one can count. He is the measurable man whose word is binding and who is

responsible. . . . he is the man of reason, the sovereign man who has become the perfect master of himself, having been tamed. . . . the man of pure expenditure who is without inhibition, restriction, or différance.[44]

DJ, like Rodrigue, also is a sovereign man of reason and self-mastery, but his sovereignty has to do with a refusal of deferral. In the end, because of Rodrigue's great self-control, he demonstrates absolute loyalty to the principles of self-mastery. Rodrigue will not submit to his internal demands for instinctual satisfaction on the level of either revenge or love: he will wait. Don Juan will not submit to external proscriptions against instinctual gratifications: he will not wait. They are complementary heroes, strong men defying equally strong forces. They represent two sides of the split that takes place in the headlong rush for and toward modernity: when the countdown toward death becomes ever more precisely measured, the subject can only learn to wait patiently with the greatest forbearance, or rush ahead recklessly in order to try to beat the clock.

Don Juan's infidelities allegorize what de Man calls literary modernity as it is determined by a temporality of precipitation and repetition. If one insists on breaking with the past, and refusing to be beholden to anything or anyone, one may find oneself only able to keep one, fatal appointment. Don Juan, after all, does not, despite everything, miss his final date with death. He keeps his word to the Statue. His faithlessness can be understood as a kind of fidelity, a *peculiar* kind to be sure, literary probably, and therefore quite unsatisfactory under what Austin might call ordinary circumstances. De Man points out that it is when literature is being most violent about making a break with history that it is being most faithful to its own specificity and, therefore, to its own past. In his reading of the seventeenth-century debate between the ancients and the moderns, de Man defends modern literature, a distinctly antiliterary tendency. In both the arguments of Fontenelle and Charles Perrault against the overestimation of ancient literary works, the justification for the legitimacy of modernity is distinctly extraliterary. (For de Man, it is Boileau, in his reactionary defense of Latin and Greek literature, who remains most faithful to "literary sensitivity." La Bruyère in his introduction to the *Characters* figures among the defenders of the ancients.) For Fontenelle, literary forms are progressively perfectible:

> In the name of *perfectibilité,* he can reduce critical norms to a set of mechanical rules and assert, with only a trace of irony, that literature progressed faster than science because the imagination obeys a smaller

number of easier rules than does reason.... Even if taken seriously, this stance would engage him in a task of interpretation closer to literature than that of Charles Perrault, who for example has to resort to the military and imperial achievements of his age to find instances of the superiority of the moderns. That such a type of modernism leads outside of literature is clear enough. The topos of anti-literary, technological man as an incarnation of modernity is recurrent among the *idées reçues* of the nineteenth century and symptomatic of the alacrity with which modernity welcomes the opportunity to abandon literature altogether.... Perrault's committed, as well as Fontenelle's detached, modernism both lead away from literary understanding.[45]

DJ is the antiliterary, technological man who has perfected a technique of literary seduction. He is the phantom of detachment who is destroyed at the moment when the struggle for literary modernism was most pitched. He is the ghost of anticipation, a ghost from the future whose threat is contained temporarily by the dramatization of the power of religion. Religion is reduced to a literary effect of modernity, deus ex machina, because Don Juan's reason is punished by superstition. Poetic justice is meted out at the end of the play, but it is only that. Don Juan is also an eminently literary hero, whose antiprogressive refusal of reasonable compromises with history makes him anachronistic and casts him out of chronological order. Technical mastery and technological progress are construed as enemies of literary temporality and interpretation, but they are also conditions of the present that make our historical position legible as one in a series of repetitious confrontations between tradition and the break with it. The figure of the mechanical principle is one that accompanies the repetitions of literary modernity.

Literature represents resistance to history, and the force of history is antiliterary in nature. Literature, however, is faithful to its own historical past when it resists most violently the force of the historical. In "Deconstruction as Criticism," Rodolphe Gasché writes: "This outside of the text, an outside that does not coincide with naive empirical or objective reality and whose exclusion does not necessarily imply the postulate of an ideal immanence of the text or the incessant reconstitution of a self-referentiality of writing, is in fact inside the text and is what limits the text's abysmal specularity."[46] Tensions produced by anachronistic formations in literary texts leave traces of historical conditions in the interpretations and readings that follow. We might venture to say that literature is that which is always untimely, and therefore that every attempt to contextualize a literary text is always an account of anachronism. In the new return to history in literary studies, there is an attempt to invent synchronicity and historical progression anew.[47] The implications of such a return to historicism allows for the canceling out or forgetting of an embarrassing debt to the literary theory of the recent past. Perhaps it was necessary to repress the insights of theoretical criticism in order to be able to relearn them, but this time with a greater awareness of the urgency of our situation. Literary theory itself was too eager to declare total victory: it believed too much in its ability to make its judgments the judgments of greater academia. Literary theory, renewed by Benjamin's material historiography, teaches us to read Don Juan's abuse of every relationship as a force that allows him to arrive at the state of emergency. Relations of exploitation are the rule rather than the exception. Difference of force, whether physical or rhetorical, is that which lays down the condition of all violence: Don Juan never shies from this insight. At the end of the play, he is destroyed by the very logic of pure force itself: his punishment is overwhelming and physical. The logic of abuse is immolated by the fires of a conscience that does not burn within. Like Benjamin's angel, Don Juan's back is turned to the future. The angel of history turns into the automaton of historical materialism when its gaze is frozen and rendered unreadable by the anthropomorphizing projections of historicism.

Chapter 6
De Man on Rousseau: The Reading Machine

MACHINAL: That which takes place with neither the intervention of the will nor the intellect, as if by a machine. Automatic, unconscious, instinctive, involuntary, thoughtless, reflexive, *un geste machinal* (a mechanical gesture), *réactions machinales* (automatic reactions).

MACHINATION: An entire system of secret schemes that are more or less treacherous. Proceedings, plot, conspiracy, intrigue, manipulation, ruse. *Ténébreuses diaboliques machinations* (shadowy, diabolical machinations). *Ourdir une machination* (to hatch a plot).

MACHINER: 1. (archaic) To form in secret designs and plots that are dishonest and illicit. To plot, to scheme, to concoct, to contrive, to conspire, to intrigue. *Machiner un complot, une trahison* (to hatch a plot or a betrayal). *Machiner la perte de quelqu'un* (to plot someone's downfall).[1]

A HERMENEUTIC DISRUPTION AUTOMATICALLY OCCURS when we read literature for its mechanical or machine-like qualities. The force of a fictional character like Don Juan, who is nothing more than a set of strategically arranged performatives, is threatening to any search for meaning. The principles of mechanical repetition and reproduction provide the conditions for a literary space shaped by the activities of so many copying machines, furiously at work. The empirical problems raised by the thematic formulation of machines *and* literature, or even the representational problems raised by machines *in* literature, are not very interesting. The machine is history, and it intervenes in the field of literary production (as it has in painting, photography, film, and philosophy) as first and foremost a challenge and a threat. Machines have been read in literature as just another thematic and hermeneutic problem in order to defer the force of such a threat. If we begin with Rousseau and find ourselves discussing Victor Tausk and Sigmund Freud, we are crisscrossing a period of time during which the pace of industrialization and mechanization intensified. The machine was already a figure of reproductive alienation: that is, under the conditions of what can be described as the nascent capitalism of the ancien régime, the machine was already doing its work, and showing the way to what would be pathologized as proliferating paranoia and alienation in the next century.

In his theorization of the literary work, Paul de Man directs us toward a reading of text or linguistic formation as machines; in so doing, he was able to continue the work of overturning traditional critical questions like "What does it mean?" for "How does it mean?" Geoffrey Bennington has commented at length on the significance of the machine and the mechanical in de Man's work, especially in terms of de Man's essay on Pascal. For Bennington, Wlad Godzich's description of listening to de Man perform his readings is also "at least in part" a description "of the performance of a machine. De Man reads like a machine. But he also reads machines; and insofar as his readings are texts, they are machines too."[2] De Man's machine-like qualities, however, are not generally seen as very positive.[3] The question "What is at work (in literature)?" is one that has led to a partial reevaluation of literary production itself. To measure the performance of a reading machine is difficult: certainly, many would be horrified at this image of a literary critic who has evolved from Bénichou's heavy-handed mechanic into an uncanny, impassive automaton. One of the reading machine's finest skills, at least in the case of de Man, is its ability to find and peel away with myopic intensity the dehiscence already at work between grammar and rhetoric.

Questions of sexual difference and sexual identity lead inevitably to an encounter with psychoanalytic theory, which inquires not after the signification of sexual difference (once again, "What does it mean?") but, rather, how sexual difference functions (which begs the question of "How does it work?") in the formation of the subject and its concomitant relationship to fantasy (and therefore reality). In the conclusion of "Lurid Figures," Neil Hertz alludes to the fact that a further engagement with the structure of pathos in de Man's work would entail an investigation of the question of sexual difference:

But we have seen that the elective embodiment of the pathos of uncertain agency is the specular structure, one that locates the subject in a vacillating relation to the flawed or dismembered or disfigured (but invariably gendered) object of its attention. Here questions of sexual difference, desire, and misogyny come back into play. They are, in de Man's writing, both recurrent and judged nevertheless to be derivative. What he liked to call "rigor" meant among other things, adopting the (necessarily unstable) position from which that judgment could be made.[4]

It is suggested that sexual difference comes into play as a part of a "specular structure": femininity is the unstated term here that seems to haunt this paragraph in spectral manner. If sexual difference is both specular and spectral, it calls for a different mode of reading—one that examines closely with greater care, if not rigor, the question of the derivative quality of "sexual difference, desire, and misogyny."

What I look for in the investigation that follows are the ways in which de Manian theory leaves open the possibility of an account of sexual difference as a determining moment in the act of reading. To find those openings, however, is to deploy a reading that is highly dependent on psychoanalytic theory. While de Man resisted psychoanalysis and psychoanalytic insights, he demonstrated the ways in which theory is more often than not formed by its resistances. It is perhaps not as perverse as it may at first appear to read de Man and psychoanalytic theory together on the question of the machine in returning to de Man on Rousseau. I propose a reconsideration or a rereading of de Man's insights into two linguistic acts, the excuse and the confession, both of which he analyzed but refused with regard to his own past. In a de Manian context, we will read Rousseau with the idea of what Freud called "the mechanism of paranoia" in mind: there will be an emphasis on the term *mechanism*, while taking into account the significance of the machine for Paul de Man's work as well. It was, after all, Rousseau who gave us an early glimpse of the infernal machine. For de Man, as well as for Rousseau, the figure of a woman, Marion, will appear in the context of various disavowals and denials. It is in de Man's reading of Marion that one finds revealed a grave inadequacy in his account of Rousseau's cover-up, and the representations of Rousseau's desire.

If *machine* seems a fairly neutral term, the shadier, more diabolical aspect of the mechanical seems to have cleaved, in French, to the term *machiner*, which describes the hatching of secret plans to bring about someone's downfall. It has to do with a betrayal, for *machiner* is to work behind the scenes while putting on what we call in English "a good face." English has inherited the term *machination,* but unfortunately, we have no verb to describe the plotting, the conspiring, and the betrayals that are encompassed by the term *machiner*. Freud and his disciple Victor Tausk point to the pathological aspect of the machine when they describe the *mechanisms* of paranoia, but the psychoanalysts refuse recourse to Cartesian differences between the human being and the machine. In psychoanalysis, the machine is anything but inhuman; the mechanism functions as a figure of psychic functioning. They are partial machines (Derrida has called them writing machines),[5] but they are not assimilated or domesticated through anthropomorphization. Once a machine is successfully constructed to receive human projections, it becomes even more uncanny. If it is a symptom of being human, then it is because it takes one (human being) to know one (machine). In Rousseau's *Reveries of a Solitary Walker,* the machine is already making a number of brief appearances in fantasies of persecution that were irradiating in all directions in his work, opportunistic and inexorable in their increasing strength.

The Influencing Machine

In his 1918 essay "The Influencing Machine," Victor Tausk elaborates on the case of a certain Nataljia A., a former student of philosophy who, deprived of hearing, could only communicate through writing:

> The patient is Miss Nataljia A., thirty-one years old, formerly a student of philosophy. She has been completely deaf for a great number of years, due to an ulcer of the ear, and can make herself understood only by means of writing. She declares that for six and a half years she has been under the influence of an electrical machine made in Berlin, although this machine's use is prohibited by the police.[6]

Tausk is vague as to why her deafness would also make her unable to communicate except "by means of writing." This raises a number of unanswerable questions for a reader of this case study. Has she always been deaf? One is to assume that she has also lost her voice with her loss of hearing. What is the significance of Nataljia's recourse to writing as the only way in which she can make herself understood? Are we to suppose (and it seems the conclusion that must be drawn) that what we have of Nataljia's descriptions of her symptoms were written rather than spoken by this analysand, this former student of philosophy?

Nataljia's disease and her symptoms are all about the unbearable distances traversed by what must have been for her an incessant scratching out of phrases and sentences that bore the responsibility of communicating with others. In describing writing as telecommunication, Derrida speculates that the written word is not only launched across a distance; it calls up the very notion of distance itself by extending the space of communication to include the places from which the writer is absent.[7] Nataljia most significantly is persecuted by a machine that is located in Berlin, which happens to be far from where she happens to reside. Tausk does not fail to emphasize the long-distance condition of her fantasies of persecution and influence. It is perhaps a lack of proximity that is part of Nataljia's problem. The machine has power over her from afar. The machine is distance itself: it both produces a separation and then proceeds to collapse that difference through a form of transmission that mimics electricity and telegraphy.[8] Transmission and distanciation are Tausk's two major theoretical and symptomatic preoccupations.[9]

Freud refused to undertake an analysis with Tausk because the disciple's theorizations of war trauma and its relationship to sexuality matched his own preoccupations too much. A disciple who follows his master too closely is in a terrible predicament. Tausk found only one way out—suicide. His suicide was accompanied by a request that his own work be destroyed: he remained faithful to the master in this final act of double self-effacement. This final demand was not fulfilled, but the record of its existence is available to us. For Paul Roazen, the psychoanalytic establishment fails to deal with the case of Tausk because it has failed to analyze Freud's own filiation and apprenticeship.[10] Diane Chauvelot also describes Tausk's suicide as the successful transmission of a constellation of speculative and theoretical difficulties for Freudians that only the Lacanians would be able to handle because they understood the problem of the paternal metaphor.[11] In any case, whether or not the Lacanians were better than the Freudians at considering the question of transmission is not a judgment I am prepared to make here.

Tausk's powerful analysis of the case of Nataljia A. continues to exert an influence on our analysis of the machine. For Chauvelot, the questions of filiation, time, space, and causality were of primary concern to Tausk: the influencing machine represents for the analyst as well as the patient a highly condensed image of spatiotemporal aberrations: "The patient does not know definitely how this machine is to be handled, neither does she know how it is connected with her; but she vaguely thinks it is by means of telepathy."[12] Nataljia's affliction can be described as a telepathology, a pathology of telecommunication.[13] Tausk argues that in cases of schizophrenia where the patient suffers from auditory and visual hallucinations as well as strange physical sensations (such as the feeling of being penetrated by electricity), the analyst should look for a machine—or more precisely, an influencing machine—from which these "foreign" sensations emanate. For the paranoid schizophrenic, the faraway machine exerts an inescapable and powerful influence as effects and affects on the body of the subject.

In the context of the intrigue or the conspiracy, the notion of machination also describes an attempt to exercise a surreptitious force in order to bring about, or "machine," someone's fall from grace. Its destructive intent is never realized as a direct confrontation: in order for a conspiracy to be completely successful, the conspirators must retain their anonymity. They must remain absent from the scene of destruction. The pernicious influence must be exerted from a distance in order to be fully realized: its coercive effects are telegraphed or telecommunicated. It was this kind of force that made Nataljia suffer.

According to Tausk, the schizophrenic's machine comes into being when genital sensations are perceived by the subject as threats to his or her regression into an infantile state. Genital sensation is projected onto the machine and redirected back against the subject as this threatening influence. The machine then becomes the origin and originator of genital sensation. This is what happens in an early stage of development of Nataljia A.'s machine: "Those who handle the machine produce a slimy substance in her nose, disgusting smells, dreams, thoughts, feelings and disturb her while she is thinking reading or writing."[14] Handling the machine produces a liquefaction in Nataljia's mucous membrane, which disturbs her. Tausk points to Freud's declaration that when machines occur in dreams, they almost always represent the genitals. Psychoanalytic theory establishes a figurative relationship between genital manipulation and mechanical manipulation, for the qualities of genital manipulation are transposed onto mechanical activity. Tausk points out, however, that often in machine dreams, the machine takes on an increasing complexity in order to distract the dreamer until nocturnal genital sensations subside. It can be understood, then, that by redirecting attention to a field of mechanical complexity, the machine works to allow the dreamer (or thinker) to defer or defuse genital excitation.

In Nataljia A.'s case, the patient declared that in Berlin the influencing machine was manipulated by her rejected suitor. This machine was the source of all her troubles. Yet it kept changing, and during the course of treatment with Tausk the machine lost much of its complexity, becoming simpler and gradually losing more of its previous characteristics. What is important here is the inherent instability of Nataljia's machine, in contrast to the machine of increasing complexity in the dream discussed above. Nataljia's machine is never quite itself: it is always changing. Strangely enough, in psychoanalysis the machine represents nothing less than the potential for metamorphosis. At first, Tausk declares that the machine was a double, predictably enough, for the patient. He continues, however, to pay close attention to Nataljia as she regresses more deeply into schizophrenia. Her relationship with her body changes in such a way that when she finally reaches a fully infantile state, she is no longer capable of differentiating between bodily sensations. The simplification of the influencing machine reflects this, as Tausk describes:

> At an earlier stage, sexual sensations were produced in her through manipulation of the genitalia of the machine; but now the machine no longer possesses any genitalia, though why or how they disappeared she cannot tell. Ever since the machine lost its genitalia, the patient has ceased to experience sexual sensations. (195)

Nataljia A. in fact becomes her genitals, or her genitals become her "self." In short, she is no longer able to differentiate between her "self" and a part of her body. According to Tausk, the machine is a crucial piece of both paranoia and schizophrenic regression:

> The evolution by distortion of the human apparatus into a machine is a projection that corresponds to the development of the pathological process which converts the ego into a diffuse sexual being, or— expressed in the language of the genital period—into a genital, a machine independent of the aims of the ego and subordinated to a foreign will. (213)

The mechanism of paranoia allows for the intensity of genital sensation to be projected onto the outside world. In the case of psychosis, machines offer themselves up as convenient sites of distorted and disfiguring projections. Machines become the mediators of self-representation in psychotic regression as a function of increasing forces of repression. Laurence Rickels discusses this in the context of the case of Schreber:

> As Freud makes clear in his study of Schreber, the world or "wealth of sublimations"[15] destroyed by the unchecked upsurge of repression and the matching massive withdrawal of libido is replaced, via projection, with another world of paranoid provenance in which media-technological attachments and controls are no longer subliminally veiled.[16]

The projection machine comes into focus as the world of sublimation is impoverished by greater forces of repression. How Nataljia is able to communicate this loss of genital sensation to Tausk is a question we will not be able to resolve; perhaps she telegraphs him from deep within her regression, the consequences of her simplifying machine. Her psychosis becomes the medium of transmission of her symptom.

Shifting gears for a moment, let us consider Rousseau and Nataljia A., who make a nice couple. Like Rousseau, she is extremely isolated from the world and suffers from terrifying fantasies of persecution. Rousseau, according to Jean Starobinski, also suffers from problems of communication. The breach that opens up in Rousseau's writing about his experience of difference from others seems to grow as the external world takes on an increasingly hostile mien:

> Before the self senses its distance from the world, it experiences its distance from others. The evil in appearances strikes first at the existence of the ego and only secondarily at the shape of the world. . . . When man's heart loses its transparency, nature turns dark and tangled. The image of the world is shaped by the way in which mind relates to mind; any alteration in that relationship distorts appearances.[17]

What happens in the mind (or "ego") alters the image of the world. This is a fair description of the general structure of paranoid projection and resembles Nataljia A.'s difficult relationship with the external world. The means of discovering this terrible distance, and the medium by which it can be overcome, lies also in writing. For Rousseau, writing is a way of avoiding the misunderstandings produced by his inability to speak in an improvisational way. To be in the presence of others is to run the risk of being misunderstood, and so Rousseau must have recourse to writing as a means of self-representation:

How can he avoid the risks of improvised speech? (What other mode of communication can he try?) In what other way can he show himself? Jean-Jacques chooses to be *absent* and to *write*. Paradoxically, he will hide in order to make himself more visible and trust to the written word.[18]

Ironically, writing is the space where the distance between the self and others becomes concrete: it opens up a nonphenomenological space of projective mechanisms and projected influence. It is both mask and veil, and the medium through which revelation and dissimulation can take place. The machine figures in both Rousseau's and Nataljia A.'s fantasies of communication as a contingency of their own fear of intersubjective and intrasubjective misunderstanding, abuse, and breakdown. But her regression into a catatonic state of self-sufficient narcissism is more successful. She is, in the end, obviously much sicker than Jean-Jacques and achieves the kind of indifference to the outside world that Rousseau hoped to accomplish through reverie and its representations. Nataljia's body undergoes hysterical conversion and becomes a genital, and then through paranoid projection it transforms itself into a machine; in Rousseau's projections, the externalization produces transformation in others. Contraband becomes a gift. The beloved becomes the object of slander, men become machines. Friends become persecutors. Truths turn into lies.

Rousseau's search for that completely self-sufficient state of reverie is narcissistically regressive, but his reveries are interrupted with greater and lesser degrees of violence. There are times when he is literally knocked out of them, as in the Great Dane accident of the Second Promenade when a dog runs over him during one of his long walks around Paris. He is knocked unconscious and upon waking is completely disoriented; he experiences a sense of having lost all notion of who and where he was.[19] Regaining consciousness and self-consciousness, he writes and produces an account of the deliciousness of this temporary amnesia. Nataljia's story is told as a case history, and the success of her regression excludes any possibility of commemoration or remembering: "Language cannot be taken for granted, and Jean-Jacques is uncomfortable whenever he must speak. He is no more master of his tongue than of his passions. What he says almost never corresponds to what he truly feels: words elude him, and he eludes his words."[20]

As Rousseau becomes more and more convinced of the malevolence and universality of the cabals and conspiracies surrounding him, he becomes convinced that he is all alone in the world. The cabal has effectively depopulated the planet of human beings outside of Jean-Jacques. Everyone else has somehow been transformed into mechanical beings with a singular motivation:

> When I saw that reason had been banished from all minds and equity from all hearts with regard to myself; when I saw an entire generation indulge itself frenetically in the blind fury of its guides against an unfortunate man who had never done, wished, or turned ill on anyone; when after having vainly searched for one man, I had to finally put out my lantern and cry out: there are no more left. And so I began to see myself alone in the world, and I realized that with regard to myself, my contemporaries were nothing more than mechanical beings whose exclusively instinctual actions I could only calculate by the laws of movement. Their behavior in relation to me could never explain whatever passion or intention I might have assumed to be in their souls. It was in this way that their inner dispositions stopped being anything for me; I saw in them only differently shaped masses, deprived of any morality with regard to myself.[21]

Animated by nothing more than the laws of mechanics, other people are deprived of all reason and all sense of equity with regard to him. It is interesting that Rousseau never ceases to emphasize the fact that the dehumanization of others is perhaps limited to their treatment of him. He is distinguished by their singular cruelty. In regard to him others behave in an unreasonable and unjust way, leaving open the possibility that they are not always and in every case quite so inhuman or quite so mad. If he shows restraint in this respect, it is only to radicalize his distinction. The conclusion of Rousseau's observations on the conspiracy is quite extreme: for him, others have entirely ceased *to be*. They can make no claim to passions of the soul, ontological substance, or internal dispositions. They have lost ontological consistency. They have become lumps of matter, shaped in different ways to resemble human beings. It is for this reason that he is alone in the world. After embarking on a futile search for a sympathetic, reasonable person, Rousseau gives up looking by turning out the lights. He will live in the obscurity of a spectral universe in which others appear to him as no more than shadows, mechanical beings, Descartes's automatons, masquerading as human beings. Mechanical beings, as we have seen earlier, are in fact exiled from authentic being. Other people are all coconspirators: as mechanical beings, they operate by the laws of a persecutory system that resemble physical laws insofar as they are inexorably deterministic. If La Bruyère's courtier was an isolated case of diabolical, mechanical being, Jean-Jacques's contemporaries have all become assimilated to such creatures; he is absolutely isolated because in his world human beings have been replaced by mechanical beings. Rousseau lives in a radically depopulated universe, a kind of body-snatcher science-fiction fantasy that can be consoling insofar as it offers a blueprint of an absolutely seamless conspiracy.

In the *Reveries of a Solitary Walker,* the buzz and hum of a diabolical machinery are like a dim background noise and disruption that, like the mutterings of a conspiracy, disturb the writer. In the Seventh Promenade, he sets up a quiet moment of reflection on a purely utilitarian attitude toward botany that motivates medicinal investigations into the area in which he is taking one of his long walks. The train of thought leads him to consider the unpleasant mineral world that is exploited and marred by industry: "Quarries, ravines, forges, furnaces, a system of anvils, hammers, smoke and fire follow the sweet images of pastoral work."[22] Thoughts such as this interrupt his peaceful contemplation of nature.

Later, lost in what he believes is a completely savage part of the mountains, a noisy machine interrupts his reverie when he is imagining himself as a kind of land-bound Christopher Columbus, all alone and temporarily safe from his persecutors, overcome by a moment of pride. At precisely this moment, he hears a clicking noise and follows it until he stumbles upon a stocking factory in full operation. Its discovery agitates him; the machine interrupts the flow of reverie and cuts short the wave of pride that has just begun to swell: "I cannot express the confused and contradictory sense of agitation that I felt in my heart upon this discovery."[23] The machine occurs here as disruption and interruption: its presence inspires agitation, confusion, contradiction. The discovery of the machine cuts through thought and contemplation, but in doing so it marks the moment for later recollection; like the Great Dane accident of the Second Promenade, the unpleasant discovery of a factory in the Swiss Alps punctuates the flow of Rousseau's thinking and allows him to recall all the more clearly what was going on at or before the moment of interruption.[24]

More Excuses

In *Allegories of Reading,* de Man demonstrates that Rousseau's texts produce a shift from the analogy of text as body to the problem of text as machine. The agency of texts is figured as machine-like: it is driven by a

De Man on Rousseau

grammatical force that not only defies authorial intentionality and subjectivity but precedes them as well. In de Man's reading of Rousseau, subjectivity becomes a rhetorical side-effect. The text contains within itself the principle of its own movement (grammar) and produces through its rhetoric the illusion of authorial subjectivity. If a text succeeds in functioning independently of human intentions, it will have become automaton—a machine containing the principle of its own movement. If we follow de Man's reading of Rousseau, we could draw the following provisional conclusions: Rousseau's text produces confessions (which in turn produce excuses) in order to produce Rousseau (as subject). The subject functions as a kind of ellipsis of textuality itself. In her commentary on de Man's work on Rousseau, Barbara Johnson summarizes what is at stake in de Man's readings:

> By locating the text-generating agency in the text's own desire to constitute itself independently of any subject, de Man sees subjectivity itself as rhetorical effect rather than a cause. If the one irreducible force at work is the machinelike grammar of textuality, this amounts, ultimately, to a definition of the subject's function in language as a potential for ellipsis.[25]

The frightening thing about such a reading is that the author can be understood as just another name for "text-generating agency." It is the text's own "desire" that keeps the rhetorical category of a subject afloat. In the case of Rousseau, the text-generating agency uses the differential forces of excuses and confessions to maintain the highest level of textual production. In Johnson's reading, desire occupies the place of the uncertain agency for de Man. The text is animated by a desire all its own: desire offers a limited, but not derivative, specularity whose rhetorical effects produce an elliptical subjectivity.

The incident for which Rousseau is still excusing himself, many years after the writing of the *Confessions,* the incident that Rousseau claims inspired his original need to "come clean," as it were, is of course constituted by the series of lies that he tells about Madame de Vercellis's pink and silver ribbon that was found in his possession after her death. By telling the story, Rousseau tries to excuse himself once again: "I did this terrible thing, but I did not mean to. I had an excuse." When the ribbon is found in Rousseau's possession, "They wanted to know from where I had gotten it. I panicked, I stuttered and finally, I said with a blush that it was Marion who gave it to me."[26] Overwhelmed by shame when asked about the ribbon, he blurts out the first name that comes to his lips. This name happens to be Marion's because Marion was very much on his mind. This stuttered proper name is the first time that Marion appears in the text of the *Confessions.* "M-M-M-Marion." But who or what is Marion, and how does such an ejaculation offer itself up to be read?

In a gesture that has to be read as provocation, de Man raises the possibility that *Marion* is actually *marion,* a noise or a "gratuitous improvisation." To concur with de Man's reading of Marion's radical emptiness, it would be necessary to overlook, however, the fact that Marion is also the one person from whom Rousseau would like to have received a ribbon and the one person for whom the ribbon has in fact been stolen. Let us recall that Rousseau's Marion was a real doll. Marion, as a variation on Marie, produces in French a diminutive, *marionette,* which was a term used in the Renaissance to describe small statues of the Virgin. Marionettes were used in puppet plays of the Passion. De Man's work on little Marion is not restricted to Rousseau. The problem of a phonetic formalism keeps de Man returning to m-a-r-i-o-n as an asemantic entity. Kleist's *Über das Marionettentheater* can be understood as doing nothing less than setting the stage for theater of diminutive *marions,* words effective at spreading and deferring guilt by association, whether as part of a Passion puppet play, or in the passionate play of a precocious servant boy.

Rousseau steals a ribbon to give to the lovely Marion, but under interrogation he blurts out a word that bears a strange resemblance to the proper name of this lovely woman: "She [Marion] was present in my thoughts; I excused myself on the first object that presented itself."[27] It would seem that some kind of pure linguistic force intervenes when Rousseau is overwhelmed by emotion. The name *Marion* is one that has already saturated Rousseau's thoughts: it is the radically contingent thing that comes from nowhere and shapes his lips with its phonemes. It is perfectly natural that this name lands on his lips when an accusatory question is asked. There is a blunt physicality to this act of stuttering or blurting; it is an untimely ejaculation. "I excused myself on the first object that presented itself." How could Rousseau really be responsible for something over which he had no control?[28]

There is another occasion on which Rousseau lies, and this other lie that he reports also occurs around women and the subject of women. He describes the incident in the Fourth Promenade. At dinner, a pregnant woman asks him a pregnant question: has he ever had any children? The pregnant question gives birth to an involuntary lie. "Blushing to my ears, I replied that I had never had the happiness. . . . They were expecting this denial; they even provoked it in order to enjoy having made me lie."[29] The blush is a physical index that points to the involuntary lie that slips through his lips. Rousseau denies that he has, even though he knows that a trap has been set for him. He imagines that everyone at dinner knows about his failed fatherhood. In fact, it seems that his questioner is the guilty one, whose complicity is shared by her other interlocutors who are all in the know, and who sadistically enjoy his discomfort. What is their excuse?

Rousseau explains why he lied about having had children:

> It is therefore certain that neither my judgment nor my will dictated my response: it was a mechanical effect *[l'effet machinal]* of my embarrassment. Before, I did not feel this embarrassment and I confessed my faults with greater frankness and honesty because I did not doubt that one would see what redeemed them. I felt at ease in myself. But a malicious eye afflicts and disconcerts me. In becoming more unhappy, I have become more timid, and I have always lied out of timidity.[30]

This kind of lie is the result of *un effet machinal*. *Machinal* describes a gesture that takes place "without the intervention of the will"; he cannot help but lie, but at the same time, because he is not the master of his actions—he is being dictated to—he is not really the one who lies.[31] This "mechanical effect" takes over and intervenes where his reason and his will fail him. A mechanical agency is pulling his strings. He veers dangerously close to the mechanical beings whom he describes in his conspiracy theory.

The lie about his children is a direct denial. The lie about Marion contains within it an implicit denial: "I did not take the ribbon—Marion did."[32] The problem, according to Rousseau, is that he only lies when the conditions for truthfulness do not exist. Lies intervene at moments when he is most under suspicion, and they are dictated to him by some unknown agency as an automatic, mechanical effect of his timidity and his sense of being persecuted. Rousseau always only lies *machinalement*: when he lies, he is most machine-like. It would seem, in fact, that he is never more innocent of the will to deceive than when he is lying. Another agency, the blunt force of linguistic material itself, has taken over: at these moments, his will is temporarily occluded by embarrassment, emotion, and so forth.

In his quest to fulfill his motto, *vitam vero impendenti*, Rousseau contemplates his lifelong service to truth and finds that it was precisely the incident with Marion that inspired in him the great horror of lying. The truth to which Rousseau has no problem confessing is that he had formed a fantasy around Marion, a fantasy in which the ribbon played no small part: "I accused her of having done what I wanted to do and of having given me the ribbon because it was my intention to give it to her."[33] When Rousseau substitutes Marion for himself, all the shifters are displaced. Instead of confessing "I stole the ribbon in order to give it to her," Rousseau implies the following accusation: "She stole the ribbon and gave it to me." In the lie, Marion replaces Rousseau as the agent and Rousseau replaces Marion as the indirect object. Uncertain agency is related to the pathos of a desirable object: it is in fact the desirability of the object that is implicated as one of its destabilizing effects.

Originally, the ribbon in Rousseau's possession is poised toward a future that will efface its past; in the future, contraband will be transformed into gift. The ribbon is waiting to be given away. In Rousseau's false accusation, it becomes weighted down by an ignominious past, the past of a theft and the past of the intended gift. The ribbon stays in place, like a hinge around which the shifters, subjects, objects, and tenses turn. The pronominal substitutions and the grammatical shift in tense take place automatically, that is, mechanically, without the intervention of Rousseau's will or reason.

For de Man, Rousseau's excuses take on the mechanical quality peculiar to linguistic production ("I excused myself on the first object that presented itself"): "By saying that the excuse is not only a fiction but also a machine one adds to the connotation of referential detachment, of gratuitous improvisation, that of the implacable repetition of the pre-ordained pattern."[34] Rousseau's lie (that is, his excuse) may be "simply gratuitous improvisation," but improvisation can also be read as displacement. The responsibility for the ribbon, its origin, and its destination are all displaced. Rousseau admits that he accused Marion of doing what he wanted to do: steal the ribbon so that it could become an exchange of the sign of love or desire between them. In accusing her of doing something that he had intended to do (giving the ribbon as a sign of desire), he makes her share if not his desire, then his guilt. When he is found with the ribbon, he is caught red-handed in his desire.

In de Man's reading of *Marion* as "gratuitous improvisation," there is no accounting for the significance of the substitution that takes place. Cynthia Chase points out that de Man shifts the emphasis from Rousseau's psychological explanations (based on a logic of shame and embarrassment) to the mechanical interpretation of the text that is based on repetition and the possibility of "gratuitous improvisation." A psychoanalytic intervention here, however, is not a psychologization. A psychogrammatical reading can be produced that offers new insight into the significance of pronoun shifting that is symptomatic of both Rousseau's text and the paranoiac enunciation. This reading for and of psychogrammar is supported by the fact that Rousseau's delusions of persecution are always on the way to language and finally achieve their most fully developed form in *Les Rêveries.* The grammatico-symptomatic structure of the Marion incident seems to lay the ground for later flowering of the paranoid system.

Rousseau's false accusation can be read as the confession to a fantasy: that Marion should be the one who offers him the ribbon. It is here that an encounter with psychoanalysis seems inevitable. The formation of paranoid projection in Freud's work describes just a complicated process of displacements and substitutions: "The mechanism of symptom formation in paranoia requires that internal perceptions—feelings—shall be replaced by external perceptions."[35] Rousseau's "internal perception" could be read as "I wanted to give her the ribbon"; this intention is projected onto the outside world as "external perception," or as "Marion gave the ribbon to me."

In the Freudian analysis of paranoia, the case of projected desire—"I love him"—becomes inverted as "He hates me." "He hates me" is a denial of loving him: it is also a sort of lie. The chain of substitutions that is designed to produce the effective repression of homosexual desire is in some sense purely grammatical. The structure of paranoiac projection is generalized in such a way that all desire if repressed is susceptible to this kind of deformation. The fact that Freud derived this theory from homosexual desires can be attributed to the higher wage of repression exacted on homosexual impulses. From another point of view, however, one could see that in attributing to paranoia a primarily homosexual structure, Freud projects himself. In the case of Nataljia A., Victor Tausk is able to use the structure of paranoid project to interpret the influencing machine that is at work in his patient's fantasies of persecution. Any desire, be it homosexual or heterosexual, when it is not tolerated by the narcissism of either a "normal" person or a schizophrenic can be projected onto an external source.

The paranoiac lives with this projection, which is directed inward as much as outward. This form of defense in paranoia leads to delusions of persecution. Another defense mechanism designed to defer homosexuality takes the form of erotomania: instead of "I love him," we find the contradiction taking the form of "I do not love him—I love her," which is then transformed into "I do not love him—I love her because she loves me." In the case of Rousseau's slander of Marion, the substitutions that take place are more homologous to the second structure of the paranoid mechanism. In disavowing one's desire, one lays the weight of that desire on others. Rousseau's slander expresses an erotomaniacal wish: "I love her so much that I have stolen a ribbon for her" is deformed as "She loves me so much that she has stolen a ribbon for me." The mechanism of both paranoid and erotomaniacal fantasies is based on a series of grammatico-linguistic substitutions.[36] The repression of homosexuality occurs in the Freudian model as a distortion and a revision of the phrase "I love him."[37]

Rousseau also denies any malicious intent in the matter of Marion. This categorical disavowal resonates like the involuntary denial that he utters when asked directly if he had had any children. In fact, the negation of malicious intent is supposed to excuse the lie itself:

> I have proceeded with integrity in [the confession] I have just made, and one will surely not be able to say that I have here tried to mitigate the blackness of my deed. But I would not fulfill the goal of this book if I did not expose at the same time my inner dispositions. . . . Never was maliciousness farther from my mind than during the cruel moment when I accused the unhappy girl. It is bizarre but true that it was my friendship for her that was the cause of it all.[38]

He would not be fulfilling what he sees as the goal of his *Confessions* if he did not make one more confession. He is compelled to expose the truth of his inner disposition, which is also the truth of his intentionality. He not only denies any intention of doing Marion harm; he says that any intent to harm was furthest from his mind at that moment. Strange as it may seem, it was his inner disposition of friendship toward the young girl that was his motivation:

> If the darkest secret of the absence of the self is kept hidden just at the moment that one speaks it aloud, this is because the rhetoric which is the key of keys surreptitiously reintroduces the authority of the self, however, deconstructive. But if this is so, the structure of rhetorical control must come unhinged, and so it does.[39]

Carol Jacobs goes on to cite de Man: "This would imply the existence of at least one lock worthy of being raped, the Self as relentless undoer of selfhood."[40] Friendship is both the key and the lock: as a semantically valid term, it offers absolute resistance to any giving up of a secret. What is undone by Rousseau's confession is the meaningfulness of the word *friendship*.

As the key to reading the undoing of the self, it is the signifier of force itself, a force of desire so strong that it causes Rousseau to impose his guilt on the innocent. To force a reading of friendship as the sign of Rousseau's violent repression implies also the forcing a psychoanalytic theorization of linguistic displacements and aberration on a de Manian interpretation. In these *Confessions,* Rousseau confesses or admits that he has done something terribly wrong but then proceeds to excuse himself of all wrongdoing by appealing to his inner dispositions that were, in fact, based on this bizarre feeling of friendship. One does not accuse one's friends of crimes that one has committed. Given the opportunity, one might accuse one's enemies of such things. With friends like Rousseau, one would probably do better with enemies.

Rousseau testifies not only to his total lack of malice, but also to his readiness for total sacrifice:

> If we were nevertheless to consult the disposition I was in when I lied, it would be obvious that the lie was only the product of an extreme shame and not at all from any intention of hurting the victim. I can swear before Heaven that at the very moment when this invincible shame tore the lie from me, I would have given all my blood with joy in order to turn the effects on me alone.[41]

Rousseau's excuses will only "work" when we "take him at his word," and it is only under such conditions that the performative that follows will make sense ("I swear before Heaven that . . ." and so on). The excuse is a place where the two forms of speech acts, the constative and the performative, confront each other. When we excuse ourselves, we are both testifying to a supposedly extralinguistic reality (our intentionality) and performing a linguistic act: "Someone's sentiments are accessible only through the medium of mimicry, or gestures that require deciphering and function as a language."[42] The excuse as constative statement cannot be confirmed through empirical investigation. In the case of Austin's example of the constative, "the cat is on the mat," we look to a mat to see if the presumed cat is indeed on it. In the case of testimony to an inner disposition, we have no such means of verification at our disposal—that is, we cannot verify that Rousseau had no intention of doing Marion harm. If we take Rousseau at his word, we must accept on oath his account of his inner disposition. There is no other form of empirical or nonlinguistic verification of the absence of malice. Because blushes signify the presence of shame and a concomitant lack of self-control that should exonerate this liar of his lies, Rousseau uses it as the external, telltale sign of his innocence.

But are blushes really sufficient? This kind of excuse that appeals to an inner feeling or disposition is that which is radically unverifiable: "Rousseau can convey his 'inner feeling' to us only if we take, as we say his word for it, whereas the evidence for his theft is, at least in theory, literally available."[43] The excuse cannot efface the theft, but the crime that it confronts is slander, a crime that is linguistic and performative in nature. When de Man describes the excuse, he insists on the fact that it is "verbal" in "its effect and in its authority: its purpose is not to state but to convince, itself an 'inner' process to which only words can bear witness" (de Man, *Allegories,* 280). Slander has to convince as well, and it too can only bear witness to its own authority. The most effective slander would be the gesture that could convince without recourse to any empirical evidence: this is the burden of the excuse as well. Excuses may at times be justified, but slander is the paradigmatic linguistic act. When effective, it exercises the threatening and tyrannical force that linguistic acts are capable of: when slander is effective, it makes the innocent appear guilty. Saying it is so makes it so. When slander is accepted as truth, the tyranny of pure persuasion realizes itself in a speech act. According to Rousseau, everyone who saw Marion could not help but love her. Her charms, however, do not prove her innocence in the fact of slander and calumny. The universal love that she supposedly is capable of inspiring fails her in the face of Rousseau's obstinate accusations.

According to de Man, the ribbon represents the possibility of reciprocity and symmetry of desire: "Reciprocity . . . , as we know from Julie, is for Rousseau the very condition of love" (283). The ribbon is the hinge on which turns the slanderous series of substitutions or displacements: Rousseau for Marion, Marion for Rousseau, stealing for receiving, taking for giving. As de Man explains, it is because the de Vercellis household is dominated by an atmosphere of "intrigue and suspicion" that "the phantasy of this symmetrical reciprocity is experienced as interdict, its figure, the ribbon, has to be stolen and the agent of this transgression has to be susceptible of being substituted" (283). From de Man, attributing importance to a "mood" is indeed surprising. Here, perhaps, we have a legible instance of the kind of specular pathos that the diffusion of agency in de Man's own text produces.

From a psychoanalytic point of view, the asymmetry of the sexual difference is always disguised on the level of fantasy, or in the fantasy of perfect reciprocity in love relations. "The phantasy of symmetrical reciprocity" is forbidden in the de Vercellis household because the very idea of love is out of bounds. Rousseau as subject of the fantasy of reciprocity is the lover par excellence. There is no evidence that his desire for Marion was reciprocated at all, so the accusation launched at Marion is also an accusation that is launched against the lack of reciprocity. (Rousseau wanted Marion to have stolen the ribbon to offer to him because he has stolen it to give to her. This fantasy tries to make up for the absence of reciprocity between them.) De Man's version implies that if the interdiction against reciprocity were not there, mutual love between the two young people would be possible. In the strictest psychoanalytic terms, however, the external interdiction against the appearance of love or reciprocity is imposed in order to conceal the inherent impossibility of reciprocation or synchronicity in the sexual relation, and in all relations of sexual difference. Love out of time and out of tune is the story always waiting to be told. The first and perhaps the most traumatic failure of love and reciprocity that takes place in that household is between Rousseau and Madame de Vercellis, whom he admires but who refuses to distinguish him from her lackeys. His admiration for her is not returned: "Madame de Vercellis never said a word to me that expressed affection, pity, benevolence.... She judged me less as what I was than what she had made me. She prevented me from appearing before her as anything else."[44] What might have most distinguished him from the others was his excessive respect for her, a gift and an offering that she refused to acknowledge. This is Rousseau's bad deal, but what about Marion?

Marion

Marion is a problem in all this, an incalculable victim whose accounts are never settled. Sandy Petrey emphasizes a failure to read "Marion" in both de Man's work and that of his critics (Steven Knapp and Walter Benn Michaels, in this case): "De Man reads the end of language's responsibility for descriptive accuracy and intentional expression as its total liberation from everything except itself. Because Rousseau 'was saying nothing at all,' for de Man he wasn't doing anything at all either. His locution had no illocutionary value."[45] Petrey goes on to show that Steven Knapp and Walter Benn Michaels do not necessarily differ with de Man on this point. De Man, however, insists on *Marion* as signifier even in the absence of intentionality:

> De Man's mistake is to think that the sound "Marion" remains a signifier even when emptied of all meaning.... De Man recognizes that the acci-

dental emission of the sound "Marion" is not a speech act..., but he fails to recognize that it's not language either. What reduces the signifier to noise and the speech act to an accident is the absence of intention.[46]

In de Man's reading of the emptiness of *Marion*, the nothingness still signifies: what it signifies concerns the nature of language as system autonomous of human intention. This leaves open a space for a psychoanalytic account of split intentionality. In their empirical system making, Knapp and Michaels are almost comically incapable of accounting for the gaps that exist in intentionality itself. They read the "absence of intentionality" as the literally insignificant. An interruption of intentionality is not pure absence, however: in the case of Rousseau, this rupture or break in intentionality, in judgment and will, signifies desire, desire for reciprocity that perhaps is at the root of his bizarre feeling of friendship for this poor young woman.

The difference between *Marion* and an indifferent, nonsignifying noise is enormous, not just because, as Petrey points out, the utterance *Marion* carries with it a heavy illocutionary weight that has immediate and disastrous consequences on the young woman who bears this name; it is also because *Marion* functions as an unconscious confession of the fact that this name is so much on Rousseau's mind. Because of Knapp and Michaels's understanding of "intention," what Rousseau said when he uttered the sound *Marion* was nothing more significant than noise, an arbitrary sound.[47] For de Man, this nonsignifying *Marion* signifies something out of its nothingness. There is a third possibility that might avoid an account of Rousseau's psychology while at the same time making something in the psychic life accountable for what is not simply a hiccup that sounds like *Marion*. It is a cry, perhaps nonsensical, but certainly significant, in a structure of displacements and substitutions.

On the question of sexual difference, it would seem that we would be wise not to take de Man at his word, for he does not address this form of difference at all. Despite the fact that Rousseau is the less trusted of the two in the household, his accusations carry a heavy weight because he is accusing a woman of initiating an amorous relationship through a gift. Stealing may be the same kind of crime for both sexes, but the giving of a ribbon is different when a woman is the giver and not the receiver. This kind of giving violates all rules of decorous behavior and *bienséance*: a young man may offer a young girl a ribbon in a gesture of gallantry, as a token of his love, but a girl cannot initiate this kind of giving. In Crébillon's master narrative of worldly initiation, *Les Égarements du coeur et de l'esprit,* Madame de Lursay does all that she can to point the way to her seduction and capitulation, without for a moment betraying that it is she who is initiating anything.[48] Madame de Merteuil in *Dangerous Liaisons* also describes in detail the lengths to which a woman must go to conceal any kind of initiative she might take in the course of a seduction.

Rousseau seems to have an inkling of the greater gravity of his accusation against the young woman. There is also an implication that having cast such grave aspersions on her character he has in fact destroyed it. This young woman was described initially by Rousseau as a "a good girl, obedient, and of a loyalty beyond all reproach."[49] In Rousseau's later fantasies of persecution, he has in fact made himself more and more like Marion, the falsely accused victim of his own slander. His persecutors, however, have no inner dispositions, at least none that he can perceive. Without inner dispositions, they will have no recourse to excuses and no chance to exonerate themselves in confessions. They are cypher-like mechanical beings whose moral depravity knows no limits.

It is quite possible to apply de Man's ideas of the radical formalism of language to the question of sexual difference, even if de Man himself seems to neglect such a reading. Sexual difference as produced in and by texts is radically formal and even grammatical. This should not lead us to conclude that there is a space of freedom beyond the grammatical confines of sexual difference where people could just be people. There is no existence beyond the parameters of sexual difference, even though such difference is inscribed as arbitrary and lexical. The radical incommensurability of the sexes is linguistically inscribed everywhere we look and read, but it seems that it is more often than not women who take the blame for this unfortunate state of affairs. De Man demonstrates that every text depends on a radically mechanical moment of grammatical autonomy to produce rhetorical effects: "The machine is like the grammar of the text when it is isolated from its rhetoric, the merely formal element without which no text can be generated. There can be no use of language which is not, within a certain perspective, thus radically formal, i.e., mechanical" (*Allegories,* 294). The purely formal aspect of a text is always marked by sexual difference, which is both arbitrary and ruthless in its hyperlinguistic formation. The grammatical machine at work in the case of Rousseau/Marion has to do with gender. If noises, hiccups, and sounds could be gendered as in the case of pronouns or proper names, then perhaps *Marion* could be at least provisionally described as such a linguistic entity. The charms of her person might in fact be communicated by the nonintentional utterance of her proper name: what might in fact be named as sexual difference itself, insofar as it is radically formal, could be grammatical and destabilizing at the same time.

No Excuses

The Fourth Promenade is, in many ways, a text primarily interested in making more excuses about the Marion episode. The meditation on the possibility of ethical lying allows Rousseau to recount the anecdotes of the young Fazy and the careless Pleince, two young playmates—friends, even—for whom Rousseau lied. These stories are told as a way of convincing the reader of Rousseau's virtue: he wants to prove that he is able to lie for the good of others. In this case, Rousseau tells lies to protect his friends, who are in fact guilty of having done him bodily harm.

De Man brings these stories of accidental mutilation into play with metaphors of writing: "Writing always includes the moment of dispossession in favor of the arbitrary power play of the signifier and from the point of view of the subject, this can only be experienced as a dismemberment, a beheading or a castration."[50] Every writer is "cut off" eventually from his or her text; every text functions independently and autonomously of its author's intentions. There is, then, always something in writing that exceeds the intervention of the will—that is, something about writing itself is of the order of the mechanical or the *machinal*. The case of Fazy is especially interesting because it involves a dangerous, mutilating machine. Rousseau and his cousin Fazy are playing in the family factory that produces printed fabrics or calicoes:

> One day I was at the drying racks of the calender room, and I looked at its alloy cylinders: their luster was beautiful to me and I was tempted to touch it with my fingers, which ran with pleasure over the surface of the cylinder, when the young Fazy, having gotten himself behind the wheel, gave it such an adept quarter-turn that it caught only the tips of my longest fingers. This was enough to crush them, and my two nails were left behind.[51]

Fazy begs Rousseau not to tell on him, and Rousseau keeps Fazy's secret so well that twenty years later no one knows why his two fingers were scarred. Two of his fingers are crushed in a place where fabric, not paper, is printed: printing fabric with patterns and printing paper with words are similar operations. In fact, there is something about pattern printing that resembles de Man's description of referential detachment in language in general and making excuses in particular—excuses that are not only fictions but also machines that produce the linguistic effect of "gratuitous improvisation, that of the implacable repetition of a preordained pattern" (*Allegories*, 294).

This is the pattern produced by machine-like operations, and it is a pattern that is printed on fabric by the machine owned by the young Fazy's family: machines produce patterns and run in the family. Now, according to de Man at least, in the chain reaction produced by testimony, it would be Fazy's turn to confess and excuse himself. Fazy may not have intended to crush Rousseau's fingers, but his operating the machine produced just such an effect. Rousseau may not have intended to hurt Marion by mechanically uttering her name in a false accusation, but his enunciation produces just such an effect. It is not clear, however, that Rousseau and Fazy are equally innocent. With Fazy, Rousseau is the victim this time, and he protects the one who has done him bodily harm by suppressing the truth.[52]

The conclusion of the Fourth Promenade is at best inconclusive. Rousseau is dissatisfied himself with his erratic system of differentiating harmless lies from damaging truths. There is something difficult about avoiding lies: the truth does not offer itself up so simply at every occasion. There are times when the truth seems impossible altogether:

> When the sterility of my conversation forced me to supplement it with innocent fiction, I was wrong, because one should not debase oneself in order to amuse others. When, carried away by the pleasure of writing, I added to real things, invented ornaments, I was even more wrong because decorating the truth with fables is in fact disfiguring it.[53]

Fictions can supplement the sterility of truthfulness and confabulations are like ornaments that decorate an austere veracity by defacing it. The truth is, then, a kind of unstable and austere backdrop that one is always tempted to decorate with a gaudy, purely invented lie. The truth does not exclude the possibility of supplementarity or ornamentation: it seems to invite it. It is always under threat of contamination. If Rousseau is trying to secure a stable entity of truth and truth telling, then he does not seem to succeed. Rousseau is aware of the failure. But what does he do but make more excuses? "Never did falseness dictate my lies, they were all the products of weakness. But this is a poor excuse. With a weak soul, one can still avoid vices. I would be arrogant and audacious to dare to claim great virtue."[54] A congenital weakness of soul, inherent to Rousseau, is the excuse for all excuses, but even this is not a very good excuse. Rousseau is able to turn all the weakness and all the mediocre excuses into something positive by giving a tour de force "performance of modesty." He concludes on a sententious and censorious note, about making exaggerated claims of virtue.

To return to Rousseau's paranoia, or what we might call his fantasies of persecution, what others have done to him is inexcusable. Freud found in the system making of paranoiacs a caricature of work of both psychoanalysts and philosophers: like the psychoanalyst or philosopher, the paranoiac builds elaborate systems out of small, insignificant details in a world superabundant and oversaturated with signs.[55] In a sense, Rousseau's philosophizing makes him a caricature of a paranoiac, an almost-too-perfect case study. The extraordinary enmity of others allows him to fully indulge in the idea of his own inimitability, his uniqueness, what we could call his megalomania.[56] It is, however, a peculiar megalomania—that of modesty, truthfulness, and good intentions, which is particularly obvious in the opening pages of the *Confessions:* "Here is the only portrait that exists and that probably will ever exist of a man, painted exactly after nature in all of its truth."[57] Rousseau becomes an excuse for the existence of his *Confessions*. Horace Walpole and others commented ironically on confessions that turn into boasts: such confessions try to secure a hyperbolic innocence. Rousseau insists on the fact that he never did harm nor did he wish harm on others: "In terms of harm, in my entire life, it never entered upon my will. I doubt that there is a man in the world who has done less harm than I."[58] Again, we can only take him at his word: as readers, we participate in the revealing of his innocence. This disavowal of ill will is the first and last of all excuses, for free of malicious intentionality, Rousseau can always justify the Marion incident. A paranoid mechanism, especially one that produces a structure of megalomania, guarantees for the paranoid subject a uniquely unassailable position.

Postscript: Inexcusability

There is no excuse for theory, either literary or psychoanalytic. Unlike other discourses, it does not purport to do anyone any good. It tries to provide material for thinking. Especially since the scandal that erupted around de Man's wartime journalism, literary theory in general and deconstruction in particular have been under suspicion of actually being bad. The crisis that this scandal created can only have a salutary effect on the thinking through of deconstruction because it has given us an opportunity to think through more clearly and less fanatically the transformations that theory has wrought in our fields of study. As Alexander Garcia Duttmann puts it: "Perhaps the task of a deconstructive thought is thus determined by the urgency of thinking its 'own' interruption (an interruption implicated in the consistency of deconstruction) without thereby falling back into a philosophy of history."[59] In short, there seems to be an increasing need for those of us in the humanities to have an excuse for what we are doing, and those of us who "do" theory seem to be most lacking in excuses. We are the least likely to appear as if we are doing anyone or anything any good. Theory exists, unfortunately, for itself: its insights are based on thinking through the complexities of both reading and writing. De Man's work especially cut off criticism from its safe, traditional preoccupations with understanding and expanding what it means to be human. His own renunciations betray a kind of intensity that his students admired and his enemies mocked: although he would not necessarily agree with an account of libidinalized writing or reading, the space that his work leaves open provides a ground on which ideas of desire can be discussed in the context of literature. Libidinalization can offer no excuses: it has no alibis. A passionate uncertainty about excuses is produced in this sort of reading, and it can be related to the urgency with which Benjamin's angel of history both announces and is powerless before a weak messianic force. De Man is a figure whose passionate detachment and renunciations inspire a great deal of ambivalence. The passionate detachment of his work is an echo of the radically dialectical take on history and aesthetics that Walter Benjamin was able to describe in his own writings.

De Man taught us that excuses only spawn more excuses. He himself made no confessions, nor did he give any excuses: his silence on the question of his wartime journalism is a bitter legacy for those who continue reading de Man. The discovery of those wartime writings did create a traumatic break in how or why we are to read de Man's work.[60] Every contemporary reading or rereading of de Man, then, proceeds and precedes in the wake of a discovery of what Derrida has described as "unforgivable."[61] When Derrida refers to the problems of de Man's autobiography by citing de Man citing Rousseau, he is all too sensitive to the fact that de Man's work on Rousseau is about the difficulties, even the ironic impossibilities, of autobiographical representation. For de Man refrained from indulging in the pleasure of self-justification that accompanies self-accusations, not out of a desire to deceive but out of his characteristic modesty. Here is how Derrida cites de Man citing Rousseau:

> The first sentence [of "Excuses (Confessions)"] announces what "political and autobiographical texts have in common" and the conclusion explains again the relations between irony and allegory so as to render an account (without ever being able to account for it sufficiently) of this: "Just as the text can never stop apologizing for the suppression of guilt that it performs, there is never enough knowledge available to account for the delusion of knowing" [*Allegories*, 300]. In the interval, between the first and last sentences, at the heart of this text which is also the last word of *Allegories of Reading*, everything is said. Or at least everything one can say about the reasons for which a totalization is impossible: ironically, allegorically, and *en abyme*. Since I cannot quote everything, I will limit myself to recalling this citation of Rousseau, in a note. The note is to a phrase that names the "nameless avengers." Nameless? Minus the crime (almost everything is there), the count is there, and it is almost correct. I mean almost the exact number of years: "If this crime can be redeemed, as I hope it may, it must be by the many misfortunes that have darkened the later part of my life, by forty years of upright and honorable behavior under difficult circumstances" [*Allegories*, 288].[62]

The forty years of righteousness and honor refer to the forty years between the slander of Marion and the writing of the *Confessions*, but in the context of Derrida's reading of de Man, the forty years of righteous and honorable living might refer, analogically, to the forty years that elapsed between 1941 and 1981. If we look at de Man's readings of the *Confessions*, we find that he was extremely skeptical about Rousseau's expiatory project. This skepticism carries with it a new weight when we consider the question of de Man's own guilt. Derrida's citations within citations confirm the possibility of reading this last chapter of *Allegories of Reading* as having been marked by autobiography, terminable and interminable. Continuing to read de Man seems to be the challenge that Derrida launches, in the wake of the discovery of the wartime journalism. The return to de Man produces, in a sense, an allegory of rereading: it is impossible to read de Man as he once was read, but under new conditions, because the going has gotten harder and the resistance stronger, the whole thing might be more interesting, more productive.

Perhaps enough time has elapsed that we can read de Man again without having to make excuses either for him or for ourselves. His own lessons can be and have been called on to help us read his almost complete silence around and about his guilt. It was he, after all, who showed us the relationship between confessions, excuses, and the question of guilt: "The excuse is a ruse which permits exposure in the name of hiding" (*Allegories,* 285). In refusing to make excuses based on testimony to "inner dispositions" ("I was overcome by the pressures of the situation" or "I didn't mean any harm"), de Man leaves us in the end with a difficult lesson about the problem of inexcusability. Neither Freud nor Nietzsche left us with many illusions about the future of inner dispositions or good intentions, but those who are sure of their good will shall always find it irresistibly tempting to indulge in the pleasure of accusing others. His work and the work of his students have inadvertently, perhaps, provided us with important lessons about the quandary of unrepresentable guilt, and the pitfalls of prolonged discipleship, the question of desire, and the resistance to psychoanalysis.

Chapter 7
Friends: Dangerous Liaisons

THE SEVENTEENTH- AND EIGHTEENTH-CENTURY NOVEL made elaborate excuses for its very existence because its fiercest critics felt that it should be abandoned as a corrupt and corrupting literary form. Georges May and later Vivienne Mylne have explored the dilemma of the ancien régime novel, and the various strategies employed to frame narrative fiction as both historically accurate and pedagogically valuable in the instruction of its readers.[1] In these novels, excuse making becomes indistinguishable from fiction making. The novel was often forced to adopt the mask of historical or archival document: it pretended to be something it was not. The duplicity of seventeenth- and eighteenth-century novelists involves the production of a structuring fiction that dissimulates the conditions of writing itself. Descartes translated the philosophical suspicion of poetry and the order of resemblance for the seventeenth and eighteenth centuries by establishing a new methodology in *Discourse on the Method*—that is, to protect the investigation of truth from the double threats of fabrication and error. Andrew Benjamin writes that Cartesian method is founded on "the necessity of the destruction of the chain of resemblances by clear and distinct perception. . . . and finally, the triumph by method over the continual threat of fiction."[2] In order to mitigate this threat, the novel has to be constructed according to double codes of propriety *(bienséance)* and believability *(vraisemblance)*, which it paradoxically produced at the same time. To quote Jean Rousset: "The novelist has a bad conscience in the eighteenth century, the novel claims not to be a novel; he invents nothing, he presents reality in the raw."[3] The negation of novelistic invention, however, is dependent on the construction of other fictions, such as the claim of a perfect reproduction of "real" chronologies and "believable" characters whose representation might serve as "inimitable models." The disavowal of fiction altogether can also be found in the frame of the novel, preface, editor's introduction, and so forth, where a highly coded narrative of the accident can be identified.

In the case of Laclos's *Dangerous Liaisons,* the writing of a novel is dissimulated as the nonintentional discovery of a correspondence. A fictitious editor presents himself as a copier and a collator; writing itself is represented as collection, collating, and copying: the authorial function is reduced to something mechanical, and authorial intentionality is given a moralizing and editorial tone. Both author and automaton come into representation as complex, highly tuned mechanisms: what they produce is a series of uncanny resemblances. They are both copying machines. The novel's line of self-defense concerns the production of a number of framing excuses. Offering excuses, as we have seen, is a way of acknowledging guilt and erasing it at the same time. The most interesting excuse offered in defense of the novel is that of usefulness—*utilité*—a category of Enlightenment thinking that soon attains the status of myth itself. In a formulation much reminiscent of Rousseau's *The New Héloïse* and the self-justifications that frame it, the claim of usefulness raised by Laclos does not fail as an excuse for his novel in the editor's preface to *Dangerous Liaisons:*

> The *usefulness* of the work, which will perhaps be even more contested, seems to me to be easier to establish. It appears to me that unveiling the means by which evil people corrupt the good is performing a service for public morality and I think that these Letters will be able to contribute something effectively to this end.[4]

An epistolary novel has a significant formal advantage in the matter of dissimulating its origins: its author can more effectively masquerade as merely an editor of letters who is making public not a work of fiction but, rather, a "real-live" correspondence that has been recovered from the lost-and-found of the archives of chance and happenstance.

It is at the movies that Laclos's *Dangerous Liaisons* is revealed as being actually less dangerous than one thinks. The simultaneous technologization and dissemination of this particular narrative in cinematic form offer nothing more threatening than another version of a romantic ideal: the Age of Reason narrative stages reason's defeat in order to rationalize the idealization of sentiment. When Janet Altman describes the epistolary novelists as having "discovered" the letter's narrative potential, she leads us toward an understanding of the novel of letters as a technological as well as formal innovation.[5] Both she and Jean Rousset compare a letter as a unit of novelistic narrative to the point-of-view shot, the smallest unit of cinematic exegesis.[6] It should come as no surprise, then, that the epistolary novel, as an early narrative-producing gadget and innovation, should help us read the dissemination of the myths of Enlightenment through the technology of mass media. Cinema copies the self-effacing and self-erasing strategies of the novel's representational apparatus and learns in the process how to tell a good story.

Laclos's epistolary novel tells one of the most primal of Enlightenment narratives: that sentiment (sensibility) and reason (sense) are opposed to one another as mortal enemies, and that the political victory of the latter is always eventually nullified by the moral triumph of the former. As Theodor Adorno and Max Horkheimer never cease to remind us, popular culture is always intellectual because its different media are constantly retelling the grand narratives of Enlightenment. Costume drama is the medium of choice for the culture industry's incessant mythmaking activities. Laclos's novel and Stephen Frears's adaptation participate for very different reasons and with very different means in the condemnation of Reason (embodied in the character of Madame de Merteuil) by representing its total victory over Sentiment (as incarnated by the Présidente Tourvel). In this struggle, however, the loser (Tourvel), as the less intelligent party, will always appear the more admirable, and it is therefore she who scores the only moral points in the deadly game. Merteuil's victory is short-lived as a bad case of smallpox and a lawsuit ruin her looks, destroy her fortune, and send her packing out of Paris; her humiliation reminds us all once again that it is never good to be too smart in matters of love and war. In Milosz Forman's film *Valmont*, the ending is radically rewritten so that Cécile, under the knowing eye of Madame de Rosemonde, savors her victory by marrying well, despite having lost her innocence. The bumbling, literal-minded goodness of Sentiment is supplemented by Madame de Rosemonde's indulgent worldliness—a form of acceptable cunning—and triumphs over Madame de Merteuil's monstrous Reason. In these two cinematic adaptations, a dangerous intelligence is evoked only to be contained.

Deadlines, Machinations, and Excuses

Adorno and Horkheimer see in Madame de Merteuil a close relative of Sade's Juliette. These two antiheroines of the Enlightenment discover that

Reason can produce its own kind of monstrous enjoyment: "Not unlike Merteuil in the *Dangerous Liaisons,* Juliette embodies (in psychological terms) neither unsublimated nor regressive libido, but intellectual pleasure in regression—*amor intellectualis diaboli,* the pleasure of attacking civilization with its own weapons."[7] Merteuil, like Sade's Juliette, represents the sadism of hyperreasonable, intellectual detachment. Laclos's heroine is a heroine of Reason: she savors the raptures of her lovers as signs of the success of her own self-control. Merteuil's attack on civilization takes place mostly as an exercise in self-discipline. Her ability to dissimulate all bodily feeling makes of her a hyper-Cartesian protagonist whose mastery of her passions is almost perfect. She is able to defer satisfactions, sexual and aggressive: her enjoyment has become purely intellectual. Her coldness is the coldness of reason's absolute dominion: she is capable of a doll-like impassivity to both physical pain and pleasure.

In Letter 81, the Marquise de Merteuil waxes autobiographical and tells the story of her sentimental education in a narrative that ends in the mastery of all sentiment. Merteuil is above all a survivor. She has to go on a fact-finding mission when left completely to her own devices with an old husband and an insatiable curiosity; she wants to learn the "facts"—of life—and in doing so she participates in an Enlightenment project, the one that is about securing

> the form of knowledge which copes most proficiently with the facts and supports the individual most effectively in the mastery of nature. Its principles are the principles of self-preservation. Immaturity is then the inability to survive. The burgher, in the successive forms of slave-owner, free entrepreneur, and administrator, is the logical subject of the Enlightenment.[8]

Merteuil writes an autobiographical narrative of which no other female character in the novel is capable; it describes her difficult and painful path to absolute self-knowledge (and self-mastery). Her ability to conquer all external manifestations of the passions makes of Merteuil the Absolute Philosopher. Merteuil's private Enlightenment has destroyed the sentimental idealization of love as superstition, but the insights that she gains are to be closely guarded. They are secrets that she cannot publish as discoveries, and yet the truths she learns through acute observation are extremely powerful. Her projects of discovery and exploration require that she develop a mask of feminine distraction and sexual insensitivity. The mask serves her in both her intellectual projects and her sexual explorations. Merteuil describes her self-education in self-observation:

> When I felt pain, I studied myself in order to achieve an air of serenity, even of gaiety; I even went so far as to inflict pain on myself purposely in order to maintain, in the meantime, an expression of pleasure. . . . I worked on myself with the greatest care and with even more trouble in order to suppress the signs of an unexpected pleasure. It was in this way that I was able to achieve the mastery of my physiognomy, which has sometimes astonished you. I was still very young and almost completely without cares, but the only thing I possessed were my thoughts, and I was outraged when someone forced them from me or discovered them against my will. (Letter 81)

Merteuil has to discipline herself in this way because according to the codes of propriety *(bienséance)*, all signs of sexual curiosity and knowledge are strictly forbidden to a young woman in her situation. Merteuil's autobiographical sketch is first of all a literary conceit, a digressive memoir within the novel that tells the story of sexual education and sexual repression.[9]

Rickels reminds us that

> from *Pamela* onward, hysteria was dictated to women by literature which thus doubled as semiological reserve of female sexuality's representation (or repression). Hysterical conversion accomplished, at the blocked or consumerist end of writing and desire, what Christianity had only talked about: the word became flesh.[10]

Stories of feminine sexual initiation are then inseparable from stories of sexual repression. Merteuil's narrative describes a theatricalized and artificial hysterical conversion that fashions itself after the models of impossible virtue and ignorance that were feminine ideals. Merteuil describes her self-conscious repression of sexual sensation as the performance that gives her power over her husband. The representation of sexual desire and enjoyment is something she learns to control through gazes, gesture, and words. Merteuil becomes her *propre ouvrage,* a sexual autodidact who has secretly rewritten a feminine education. Yet the publication and dissemination of her pedagogical methods would destroy her power. Hers is a knowledge that cannot be shared with other women; it can only be used against them. Merteuil puts on masks of immobility so that her face never expresses her experience. She is the ultimate unnatural woman. What

Merteuil learns is to adopt the posture of the most famous of hysterical symptoms: frigidity. This is her grand achievement and her greatest weapon. With the adoption of this symptom, she is able to better observe the world around her. As Absolute Philosopher who has discovered a secret method of arriving at truth, she is also the ultimate hysteric who has accomplished a hysterical conversion on her own body and made herself utterly unreadable to others.

What does she know that the other women of the novel do not know? She knows how to play with words *and* she knows something about sexual enjoyment and the way that it shapes sexual difference. The coincidence of two orders of knowledge—one sexual, the other linguistic—is not serendipitous. When Merteuil writes to Cécile that one should never write what one thinks, she has learned this imperative from her sexual lessons: a woman should always know, but never show how she feels; there can be no transparent, linguistic representation of the self—only differently modulated versions of it, presented for the other's enjoyment. Her letters are primarily strategic and performative, whereas Tourvel's correspondence aspires toward expression and communication.

Merteuil wants to know about pleasure, and in her explorations of her own reactions and experiences she treats herself like an automaton: she studies her own reactions with extreme care, and she works on herself accordingly in order to develop a mode of behavior that allows her to obey the constraints of being and becoming a woman. In doing so, she theorizes femininity by continuing to think through the vicissitudes of love and pleasure. What she has succeeded in (re)producing is a new and improved version of herself: a perfect version of (feminine) Reason, whose pleasures are entirely unprincipled but perfectly dissimulated. Adorno and Horkheimer call Odysseus the first figure of cunning, who participates in an early Myth of Enlightenment by outwitting the prehistoric gods with his finely honed survival skills and do-it-yourself know-how. He is infinitely adaptable and very rational. Merteuil is a latter-day, feminized Odysseus who performs one of the final phases of modernization by freeing herself from archaic notions of feminine sexuality. In so doing, she transforms herself into a cipher: "I can say that I am my own creation" (Letter 81). She is both evil scientist and monster in the same person.

Merteuil claims never to have been influenced by a desire for enjoyment: she has only wanted knowledge. In the absence of a sex education course or teacher, the desire for knowledge provided her with the instruments of self-instruction ("the desire to learn provided me with the means to do so"). Starting from a point of total ignorance on matters of love, it is mostly an intellectual curiosity that drives her to seek answers about the truth of relations between men and women. Freud has suggested that the inhibition of sexual curiosity in children often leaves them intellectually timid. Similarly, Merteuil understands the pursuit of the truth as intimately related to the mastery of erotic life. First, she naively asks her Confessor about the nature of love, only to be violently rebuffed, left no wiser, except convinced by his severity that its pleasures must be great. Married to a much older man, she feigns frigidity and stupidity with her husband. At first she is caught up in the whirlwind of social life in Paris before being taken to the country. In her husband's country house, she begins a reading program, which, accompanied by her recent experiences, convinces her of one thing: "It was there that I became convinced that love, vaunted as the cause of our pleasures, is only the pretext for them" (Letter 81). The reward for her cool-headedness is a powerful aphorism that apparently immunizes her against the suffering of passion, and the weak-minded superstitiousness of her sex.

Having understood the mechanics of sexual pleasure and its detachability from sentiment, Merteuil concludes that pleasure precedes love: the former must find a pretext or an excuse for itself by associating itself with the latter. Love, then, is a state of delusion, and it does not last: all lovers are in some way interchangeable. The pleasure principle is a mechanical one, and an intelligent and reasonable person can learn to objectify his or her own pleasures. This is a cruel reversal of Rousseau's *effet machinal,* which is produced by a loss of mastery. Pleasure can be produced like a mechanical effect. The image of the machine appears when there is a disjunction between a locution and an "inner disposition." Rousseau uses these moments of radical disjunction as excuses; Merteuil instrumentalizes them and demonstrates in her autobiographical narrative that when the speaker or writer is capable of consciously manipulating the disjunction between the psychic motivation and the linguistic utterance, the machinations becomes possible. Merteuil's machinations have to do with sexual enjoyment and sexual identity: what she machinates most effectively is the construction of her self. Merteuil is a figure of autoengineered hypervigilance.

The figure of the machine and machination haunts the criticism of the novel. For Aram Vartanian, Merteuil's character is monstrous in her ability to machinate because she represents a threat to the stability of sexual differences: she has a man's head on a woman's body.[11] This monstrosity is constructed from her "*méchanceté,* her machinations, and her sensuality." Yet when Vartanian writes about her, he seems to be condemning her with the fascination and horror that he attributes to her male contemporaries and readers, rather than performing any sustained literary analysis. Merteuil is prepared for the production of both fascination and horror: what she is capable of thinking through is precisely the necessity of feminine masquerade and dissimulation in the exercise of her intelligence and reason. She has to cover up her discovery of the imperative to cover up. In this way, her talent for dissimulation is one that she shares with the author of the novel: "In the *dénouement* that follows, the proper role of each side is re-affirmed with a vengeance."[12] Her punishment at the end of the novel not only reinstalls the primacy of sexual difference; it also stages the destruction of a copying machine of satanic intelligence. Vartanian's reading of Merteuil's struggle as one that is only mediated by sexual difference limits our understanding of both her triumphs and her defeat and does nothing but strengthen the force of her punishment.

Peter Brooks uses the metaphor of mechanization to describe processes of "dehumanization" through erotic reduction: social relations become erotic relations, human behavior becomes erotic comportment.[13] For Brooks, Merteuil (and Valmont) have mastered the codes of behavior because they are the best readers of earlier novels of worldliness. Once social laws as they have been represented in novels are completely mastered, they can be manipulated: Brooks compares Valmont's and Merteuil's work of reductive manipulation with mechanization. The novel itself becomes the manual of worldliness, giving precise instructions by which a perfect mechanism of erotic domination and reduction can be launched. In short, the libertine's refinement of the codes of worldliness as mechanisms of erotic domination is a rewriting of the earlier novels of worldliness. In *Dangerous Liaisons,* their various strategies of domination and deception are coordinated like so many mechanisms to operate in tandem as different parts of an infernal machine. The machine in its infernal ability to copy, reduce, simplify, and master is produced by the intelligence of the libertine conspirators.

If Brooks uses mechanization to criticize the libertine instrumentalization of human relations, Valmont and Merteuil use the image of the machine in order to paint a picture of stupidity. For Brooks, as for Valmont and Merteuil, the machine represents radical simplification. In a letter to Merteuil, Valmont characterizes the company at Madame de Rosemonde's chateau as automatons. He is in a rage because he has discovered that the

Présidente has returned abruptly to Paris without his knowledge. He consoles himself with the thought of Merteuil's friendship:

> I have more than once felt how useful your friendship could be; I feel it now at this moment because I feel calmer when writing to you. At least I am speaking to someone who understands me, and not to the *automatons* around whom I have been languishing since this morning. In truth, the longer I live, the more I am tempted to believe that the only people worth anything in this world are you and I.[14]

In this particular context, ignorance makes the others seem like automatons to Valmont: only his *friend*, the Marquise, can be of use to him. What distinguishes Valmont and Merteuil from the automatons are their shared understanding and intelligence, which function on both linguistic and sexual levels. To cast aspersions on the machine is to forge stronger bonds with other human beings. An automaton is both incapable of thought or expression: to be a human being among automatons is to be in desperate need of a friend.

If Valmont addresses himself to Merteuil as the one who understands, it is because she, like him, machinates. Tourvel does not. Her sudden, nocturnal flight from Madame de Rosemonde's is, like all of her gestures, expressive. It is not so much a ruse as a sign of her desperation: she is hiding nothing in her precipitous return to Paris. She also only reads expressively and takes everything literally; she writes without consciousness that there may be a difference between what her writing "means" and what her writing does. If Valmont and Merteuil are libertines, it is because they play with language in a way that none of the other letter writers of the novel does; this is of course most obvious in the scandalous Letter 48 written by Valmont on Emilie's back. The double entendres in this letter are meant to insult Tourvel's literal readings; her complete inability to read between the lines or perceive that words and phrases can have double meanings leaves her in utter ignorance of the fact that language can be manipulated like a machine when one writes things one does not "think" or "mean." What we could call her obtuseness, her stupidity, allows her to get it right in the end,[15] because even though she is unable to read Letter 141 as merely a copy of a formulaic breakup letter, she does see that it marks the end, or at least an end. In Austin's terms, she does not "do things with words."[16] Words only do things to her, but her misreadings pave the road to her sublime martyrdom to love.

Tourvel insists on receiving the letter as an authentic document, even if

Valmont tries to take it back. Tourvel's inability to machinate or see through the machinations of others has something to do with her virtue. Merteuil, on the other hand, is the one who plays with language and appearances. More specifically, she plays with all three of Austin's exemplary performative speech acts: making a bet, contracting a marriage, and declaring war.[17] Merteuil places a bet with Valmont and offers herself as the reward. She later declares war on him. She does not have the power to marry or unmarry anyone, but she does conspire to make a mockery of Gercourt and Cécile's marriage. When she takes action, it is also calculated to produce an effect of one sort or another; she has perfected the manipulation of perlocutionary effects of highly conventional speech acts. Her primary activity can be said to be linguistic. She is always reading and writing. She writes to Valmont in order to enlist him in her machinations; she writes perfectly phrased letters of deceit to Madame de Volanges to influence her decisions about her daughter.

It is between Valmont's cry to his friend (Letter 100) and Merteuil's reply (Letter 111) that one part of her machinations is revealed. She is intent on securing revenge on Gercourt, her former lover and Cécile's fiancé. When Madame de Volanges finds herself worrying about her daughter and wanting to soften her position on the question of her marriage, even contemplating letting Cécile marry the one she loves, Merteuil preaches about austere virtue and advises her to remain firm in her position. Merteuil then writes to Cécile a letter in which she mixes truths and falsehood into a powerful concoction of pure manipulation. The letter is a lesson in deception: Merteuil wants to teach Cécile what it is to machinate. She advises her, first of all, to continue her relationship with Valmont because it will help her conceal her love for Danceny from her mother. It will also allow her to satisfy her (sexual) curiosity while appearing virtuous to her lover. Merteuil tells Cécile that her mother is hoping to trap her into a confession of her love for Danceny and advises her to lie. As a postscript, Merteuil gives Cécile some affectionate advice about writing that outlines the basic strategy of machination: never write what one is thinking, always write what the other wants to hear, especially when the addressee is one's lover. This is the description of the loss of innocence—in language. In this narrative, linguistic initiation comes after sexual initiation: the latter only serves to pave the way for the former. Playing with language is associated with libertinage, moral turpitude. Puns, double entendres, and linguistic games are akin to sexual manipulation and erotic mastery. Even as all human relationships are increasingly rationalized and instrumentalized, the power of love and sentiment must be all the more rhetorically exalted and idealized.

Merteuil wants not only to make Cécile unhappy (by insisting on her marriage to Gercourt); she wants to exploit this unhappiness by turning the girl into an agent of intrigue. When she decides that the girl is corruptible not out of intelligence but out of stupidity, she condemns her in the following manner. Her judgment is cruel, not because it is exaggerated but because it is only too accurate in its assessment of Cécile's weakness, which is above all

> a weakness of character that is almost always incurable and which is opposed to everything in such a way that while we thought we were preparing this little girl for intrigue, we would only be making of her an easy woman. I know nothing as uninteresting as the easiness that comes from stupidity, that makes a woman capitulate without understanding how or why, but only because one attacks and she does not know how to resist. These kinds of women are nothing but pleasure machines. You will tell me that we might as well make the best of it, that that is enough for our plans. . . . but let us not forget that everyone learns the springs and motors of these machines. Therefore, in order to use this one without risk, we have to hurry, to stop early, and then to destroy it.[18]

The pupil has proven unworthy of the master: she is not an initiate, she is a machine, and the pleasures she procures and produces are only mechanical. Her destruction will take place at the hands of her husband. Marriage spells the destruction of both pleasure and machine. Merteuil gives Cécile's sex life a deadline. Love does not last, but neither does pleasure: this is the conclusion she has drawn. This particularly brutal assessment of Cécile reflects Merteuil's absolutely reasonable calculations.

Merteuil can be honest about Cécile with Valmont because he is her friend and coconspirator. They are brought together by common interests and common insights, a deep complicity that is not or is no longer erotic. Mutual respect for the other's skills at intrigue and seduction is their bond. When Valmont wants to be more than friends, Merteuil places a number of conditions on the renewal of their amorous relations. First, she makes the successful seduction of Tourvel a requirement of their reunion. Then she adds Cécile's corruption to the terms of this contract. When Merteuil gets angry at Valmont for describing Tourvel and Cécile in terms a bit too glowing for her taste, he defends his fidelity to the emptiness of conventional forms in his formulations about the two women. If he dispenses with such forms in his correspondence with the Marquise, it is because of the confidence and trust on which their friendship has been built. As their relationship disintegrates, however, it becomes clearer that despite the emptiness of forms, they are essential in liaisons of any kind. Valmont forgets to treat Merteuil "like a lady." Responding to her anger that he has assumed her consent before attaining it, he responds, "I know very well that custom has introduced in this case a respectful doubt; but you know that it is only a form, a simple protocol. It seems to me that I was authorized to believe that such exacting attention to custom was no longer necessary between us."[19] So careful about matters of form and protocol in other places—in the seduction of Tourvel, for instance—he "forgets" in the name of friendship to take such precautions with Merteuil. His lapse is fatal here because it leads to his failure to seduce Merteuil, and the subsequent declaration of all-out hostility between them.

After seducing Tourvel and corrupting Cécile, Valmont insists on his reward. When pressed, Merteuil seems to offer him an escape clause from their contract: "Let's just be friends" (Ne soyons qu'amis). Valmont balks at the idea. Friendship is a tricky business between the two of them. Immediately following this offer of friendship, Merteuil presents what seems at first glance to be a conjectural description of the sacrifice that she would demand of Valmont, if they were to become more than friends. What she demands is that Valmont break with Tourvel. In the classic fashion of the lady of courtly love, she demands of her knight an impossible task that must be accomplished before her favors are made available.

Valmont thinks that this demand is a simple one to fulfill because he does not believe he loves Tourvel. He treats her badly and sleeps with Emilie, his erstwhile writing table. He offers these actions as the signs of his lack of love to the exacting Merteuil. "I insist, my lovely friend, that I am not in love. *It is not my fault* if circumstances force me to play the role."[20] It is not his fault if he has to play the role of besotted lover with others. He would not deceive his friend. Merteuil and Valmont are never being more radically formal and dissimulating than when they are negotiating the terms of their friendship; it is when Valmont tries to efface the importance of forms between friends in his effort to have Merteuil as a lover again that a hostile confrontation is precipitated. For the first part of the novel, they keep the question of their relationship at bay because they act as conspirators whose common interest unites them. They conspire with each other and with the purely formal, gratuitous effects of the rituals of seduction. Conspiracy functions as the effective condition of alliance in the absence of more idealized bonds. When the question of love between them is raised, however, so is the possibility of enmity.

When Merteuil choreographs the breakup of Valmont and Tourvel, she does so by offering him an opportunity to use his excuse on another woman. She has found in Valmont's "It is not my fault" a sign of his aggressive unaccountability with women, and she decides to use this weapon against another woman, a woman with whom he may very well be in love. She tells the story of a woman friend who helps out a man of the world who has found himself in a bind. The man has made himself ridiculous over an unworthy woman. He must be given an out, an excuse for extricating himself from such a situation. In this first instance, an excuse is given as a gift from a friend to a friend. Merteuil's gift is a breakup letter that supplies the excuse of all excuses: "It is not my fault." The giver, however, is only citing Valmont himself, whose motto was, in the face of his actions, "It was not my fault." This is, in a sense, the *original* excuse. This formulation is always held in reserve by the woman who foresees, in the "application" of the remedy, that something can go awry. She can always then use the excuse on him, this time performatively, giving it not as a gift but as an act of self-exoneration. She has included no remark with the body of the letter, nor has she signed it. She implicitly and covertly cites his no-fault signature excuse as an insurance policy against all responsibility, by not offering an introduction to or imperative in the letter. In this way, she has recourse to "It was not my fault either." The absence of commentary and signature seals an insurance policy against blame or responsibility. The ambiguity of the letter produces an escape clause for its author.

Valmont reads the story as an imperative for replication and dutifully copies "It is not my fault" and sends this missive to Tourvel, not realizing that Merteuil has turned his own words against Tourvel and himself. He only hopes to profit from his submission to Merteuil's law of duplication and dissimulation. What he forgets, of course, is that Merteuil was the original recipient of Valmont's "It is not my fault"; she reroutes the insult to her rival. Valmont cites himself by way of Merteuil without even realizing it. He thus allows Merteuil to write his memoirs and script his life, at least for a moment; she does so by dictating to him an insulting letter, using a turn of phrase that he has tried to use on her. Merteuil has calculated the exact measure of his overreading by playing to his overestimation of his own sentimental detachment. He is effectively remote-controlled by

Merteuil and sends a copy of the template of all breakup letters to his lover, Tourvel. Many of the epistles in this novel have devastating effects, but few are as cruel as the one that Valmont copies from Merteuil's missive. Here is the story that introduces the letter:

> A man of my acquaintance entangled himself, as you did, with a woman who did him little credit. He had the good sense, from time to time, to feel that sooner or later this affair must reflect adversely upon him.... His embarrassment was all the greater for his having boasted to his friends that he was absolutely free.... he spent his life committing one stupidity after another, never failing to say afterward, "It was not my fault." A woman, more generous than malicious by nature, who was a friend of this man's, decided to try a last resource to help him; but she wanted to be in a position to say, like her friend, "It was not my fault." She therefore sent him as a remedy, the application of which might be efficacious in his illness, but without further remark, the following letter. (Letter 137)

Valmont merely reports to Merteuil that he has sent the cruel missive, and Tourvel reports to Madame de Rosemonde that she has received it.[21] The letter only appears in the novel as a "letter within a letter," as citation. "It was not my fault" punctuates the sententious and cynical statements on love's vicissitudes. It is in the order of things that love does not last: one cannot reasonably expect it to. Therefore, to fall out of love is no one's fault; it is only natural. The message of the letter is a didactic one: every love affair cannot escape its deadline. In view of this law of nature, it is not *reasonable* to expect love to last. This is the lesson that Merteuil wants to teach (Tourvel), the lesson that she has taught herself.

The letter is a citation of a citation, but its status as copy is not legible to its addressee. Upon its reception, Tourvel forwards it to her confidante, Madame de Rosemonde and renounces her role in any correspondence whatsoever. Valmont in a sense was only forwarding Merteuil's message to Tourvel: he takes on the role of postal agent in order to do his friend Merteuil a favor. What is an excuse between friends? It is a gift that both effaces blame and nullifies responsibility. If excuses are empty forms, signifying nothing more than respect for convention, they are nonetheless not to be neglected. In neglecting them with Merteuil, Valmont takes the first (unconscious) step in declaring that the impossibility of their love will lead to a declaration of hostility. Forgetting protocol is assuming that there is some substantive, unmediated relationship possible: Valmont seems to be having a Rousseauean moment when he argues for the non-essential nature of empty forms and *précautions minutieuses*. When making excuses to Merteuil about having to a play at being a lover to other women, he seems to believe in the possibility of a relationship stripped bare of form or protocol, those ornamental, supplementary precautions and pretexts. He takes his excuse making lightly, but Merteuil does not.

Making excuses implies the simultaneous acceptance of guilt and the refusal of responsibility: the most effective excuse would be the one that completely erased the conditions of its own necessity. What was most striking about Rousseau's account of the case of the purloined ribbon, however, is the attempt to efface all blame through excuses. "I maligned an innocent girl, I have suffered for this terrible lie my whole life—but it wasn't my fault. I meant her no harm." The ideal excuse is the one that has recourse to a universal law of nature, that reason has identified both as an a priori and a final recourse—that is, "Love never lasts." The limited half-life of love is the excuse for every vagary of the heart, but this excuse is a pure fiction even within the frame of the novel, for it is evident to Merteuil (and perhaps even the reader) that Valmont's passion is not at all spent. As she writes to him: "One is very soon bored with everything, my angel; it is a law of nature. It is not my fault.... I quite realize that this is the perfect opportunity to accuse me of perjury: but if, where nature has gifted men with no more than constancy, she has given women obstinacy, it is not my fault" (Letter 141).

When de Man writes that the excuse is always a fiction—and not just any fiction but a fiction that produces and reproduces preordained patterns like a machine—he describes the way in which Valmont turns himself into a copying mechanism. The formula of disavowal is not even his own.[22] "It was not my fault" produces a number of aphoristic formulations about the relations between men and women, but the sententiousness of Valmont's ghost-written letter is a powerful fiction that lays down for all lovers and passions an inescapable deadline, a date after which all romance goes stale. When Valmont copies and forwards the form letter, he believes too much in his own mastery of the purely formal aspects of the postal system. He thinks that by filling in Tourvel's name in the place marked addressee, he can reduce her in such an exchange to just one more address on a mailing list of rejected lovers. He and Merteuil are great masters of "referential detachment" and "gratuitous improvisation," but Valmont lets the power of the mail-merge function go to his head. The end of this romance only proves the power of Sentiment. Merteuil knows that Valmont is in love with Tourvel, but she also knows that he is too proud or too stupid to admit it. Her prescience, however, does not account for one thing: the retribution of others. For even if Valmont and Tourvel are destroyed, their love for one another is not: the proof of its durability lies in Valmont's last acts of self-exposure and the betrayal of Merteuil to Danceny. Only in such an act does his love for Tourvel prove itself: it is ironic, then, that only when love's force is turned toward destructive vengeance does it triumph over Reason. The eclipse of Merteuil's reason takes place as love's Pyrrhic victory.

In Frears's *Dangerous Liaisons*, it seems quite clear that Valmont is in love with Tourvel at the end of the film. In the role of Valmont, John Malkovich plays the dying scene with as much sincerity as he can muster up. During the duel, he grows increasingly listless and fights with less and less enthusiasm. He seems to allow Danceny to wound him, and he aggravates the wound, making it fatal, by thrusting the sword in deeper as his opponent looks on in horror. With his dying breath, he entrusts Danceny with a message to Tourvel. Valmont is transformed from cad to lover by his martyrdom. When Danceny delivers the message to the delirious Tourvel, he whispers it to her, and the suffering is effaced from Michelle Pfeiffer's features: death comes to her character as a relief from uncertainty.

In the novel, however it is much more ambiguous. Tourvel dies in the convent after having refused all letters from the outside world: there is no delivery of a final message. This ending is much more unsettling epistemologically, and in turn, more haunted by hermeneutic uncertainty: is Valmont's transformation into the lover just one more act? All forms of knowledge about his inner disposition seem tenuous. Only Merteuil seems to "know" with some certainty, has read Valmont with any amount of confidence. Merteuil declares that Valmont was so clever that he was able to fool himself; by writing to the Présidente what he thought were mere lies about his love for her, he was actually telling her the truth. Being so accustomed to lying *automatically,* however, in the course of seduction, Valmont takes his own lies and suppressions of the truth at face value; he forgets that the truth can often appear as deception. He thinks that he is the master of his own truth, but his deception turns into self-deception.

Frears's film emphasizes Valmont's defeat as a mystical transfiguration. In order to strengthen the "cult of feeling," a hyperreasonable position of total detachment must be produced as its monstrous opposite. When Adorno and Horkheimer describe the romantic hero of cinema as an important figure of Enlightenment, they explain Valmont's metamorphosis at the end of the film:

> The exuberantly tender affection of the lover in the movies strikes a blow against the unmoved theory—a blow continued in that sentimental polemic against thought which presents itself as an attack upon injustice. Though feelings are raised in this way to the level of an ideology, they continue to be despised in reality.[23]

Injustice here is embodied not by the social order that forces women to pretend that they are stupid and frigid: injustice is incarnated as the force of Merteuil's machinations against the lovers. The Enlightenment produces sentimentality as an ideological by-product whereby the loser as lover is always idealized. Sentimentality is a kind of violence that is actually sustained ironically by contempt for the stupidity of feelings. The novel proves itself most useful in the dissemination of the fictions that will nourish such cults of feeling, well into our own century. What the cult of feeling allows for is the exclusion of violence, exploitation, and abuse from the precincts of love. When the abuser is punished, the ethical balance of love can be made right. This is obviously a strategy by which the threat of deception and aggression is anthropomorphized and demonized. Merteuil is right, after all, that romantic love is a superstition, but the revenge that she tries to exact for this insight is purely personal. The colder and more reasonable she is, the more passionate and confused are the lovers. This is the dialectic of love and reason.

The cult of sentimentality is founded on the idealization of romantic relations between the sexes. This process of idealization seems to have intensified its efforts on behalf of benighted innocence, even as women, in industrialized societies, have gained greater autonomy and self-determination. It seems that it is precisely in the world of "liberated" and "modern" women that the cult of sentimentality is the strongest: without the heaviness of traditional patriarchal taboos on feminine behavior, it is love itself that must be exalted so that a woman does not go too far. If Tourvel seems confused about love and friendship, she is being no more than a literal reader of Rousseau, while Valmont and Merteuil are better readers of her reading. When Rousseau confronts in his recounting of past events the slander of Marion, he offers as an excuse a feeling of bizarre friendship for the object of his desire. When confronted by Valmont's declarations of love, Tourvel offers friendship and respect in return. Both the naïveté of the young Jean-Jacques and the cynicism of Valmont produce friendship as an excuse for a relation of force with regard to women. Valmont avails himself of the terms of friendship, while understanding them not so much as empty but as pretexts, or excuses, for the pursuit of other things (which fall in the realm of the sexual or erotic conquest). As Tourvel is forced to receive his letters, Marion is forced to receive Jean-Jacques's guilt.

Friends

The problem, of course, is that the women in question, Tourvel and Marion, are both surprised by the outcome of what Derrida calls "heterosexual" friendship.[24] In Merteuil's hyperreasonable system, she discovers that talk of love is nothing but a pretext for sexual enjoyment. Pretexts are notoriously empty of meaning: they are always linguistic placeholders for the deferral of other terms. The pretext can be compared to both the excuse and the lie, but its complicitous relationship with socially acceptable forms of behavior produces a haze of ethical obfuscations around its usage. From the absolutely demystified perspective of a Merteuil, the pretext of friendship between men and women is always a cover-up for pursuing a sexual or erotic relationship in which seduction and conquest produce the double pleasures of victory and capitulation. When Merteuil describes Prévan's courtship to Valmont, she dismisses his discussion of "delicate friendship" as an empty conventionality, a "banal banner" under which they embark on their campaign against each other (Letter 135). The military metaphor is interesting insofar as the feminine surrender to pleasure is always figured as a masculine victory. Giving in or giving it up brings a certain pleasure for the woman, but the fantasy of male domination in matters sexual must at all costs be maintained by the idealization of women. Rape is the logical outcome of such relations, and the accusations of rape launched against the hapless Prévan are so believable because they support that fantasy.

While Valmont and Merteuil may see the banal banner of friendship as merely a pretext for other kinds of less-than-respectful relations between the sexes, they do have an enormous respect for the power struggles that take place in its name. When Valmont replies in categorical terms that he will not be Tourvel's friend, he understands that what he is engaged in is an important "conflict of words" *(dispute de mots)* that will move him closer to the victory he desires. He explains to Merteuil that he has strategically refused the title of friend by insisting on another title, that of lover. As we have seen, what he and Merteuil share is a special knowledge: they know how to produce affects that are completely divorced from psychic motivations or inner dispositions. Like Don Juan, Valmont and Merteuil believe in the radically formal use of the conventions of seduction. If all three of these literary figures must be destroyed, it is because the victories of intelligence in matters of love must always be temporary ones. It is as if romantic love analyzed too closely will be transformed and reveal itself as being no more than a pretext for relations where erotic domination seeks an alibi and instrumentalization of the other is masked as harmonious reciprocity.

When Valmont declares his love for Tourvel, she replies that she "just wants to be friends."[25] Once again, friendship is proven to be nothing but a pretext that allows him to appeal to Tourvel for more opportunities to speak to her of his love. With Merteuil, he shares his insight that "a woman who consents to speak of love soon ends up feeling it."[26] Tourvel's pleas for understanding inspire admiration and then contempt; her capacity for feeling coupled with her blindness for strategy makes her the perfect victim:

> Abandon this language that I neither can nor want to hear: renounce this feeling that both offends and frightens me. . . . Is this feeling the only one that you can know, and will your love be even more guilty in my eyes in its exclusion of friendship? . . . In offering you my friendship, Monsieur, I give you everything I have, everything of which I can dispose. What more can you desire? In order to give myself over to such a sweet feeling, I await your promise and word that I demand of you: that this friendship will be enough for your happiness. If you are as you say, corrected of your errors, would you not prefer to be the object of friendship with an honest woman rather than the source of remorse for a guilty one?[27]

Tourvel's language betrays her at every term: even in asking Valmont to abandon the rhetoric of love, she indulges in the pleasures of confusion. She wants to give him everything so that she can enjoy the sweet feeling that should be enough for his happiness. The terms of her description are highly eroticized, but her belief in the power of respect and distance that friendship guarantees masks her own desire for a kind of "safe" sex, in which sweetness can be found and everything given without risk of contamination. Her rhetorical naïveté is mirrored by Cécile's confusion over her feelings for Danceny. There is an answer to Tourvel's rhetorical final question, and it is not in the negative: in Valmont's libertine logic, he would, in fact, prefer to be an object of remorse for a guilty woman because he is at the very least highly skeptical of and not very susceptible to the discourse of sentimentalized friendship. The consistency of Valmont's discourse lies in its inconsistency; what remains constant is his absolute, hard-edged cynicism about absolute meanings. Valmont machinates his seduction of Tourvel using the linguistic confusion between *amitié* and *amour*: he refuses the title of friend only to demand the rights of friendship when he sees Tourvel avoiding him: "What have I done to lose this precious friendship, of which you have, doubtless, considered me worthy because you offered it to me once? . . . In fact, was it not in my friend's bosom that I deposited the secret of my heart?"[28]

Similarly, the *délicate amitié* that Prévan offers up to Merteuil is a formality. It refers to a cool mutual respect for conventionality, for the ossified codes of *bienséance*. Merteuil thus views the friendship that Prévan offers her with irony, but she does nothing to contradict him: she allows him to believe that she believes in the sublimity of such an idea when she sees it as no more than a rubric under which the skirmishes of a campaign of seduction can be initiated. Friendship, then, refers most of the time to pure convention, but the category of friendship is especially empty in the case of women. This is nowhere more obvious than in Valmont's recounting of Prévan's triumph over "the inseparables." These three beautiful young women were objects of admiration because they seemed in the eyes of the public to enjoy the perfect friendship. Their friendship is perceived as being so powerful that it overshadows all other relationships in their lives. What Prévan demonstrates through his elaborate machinations, his plots hatched in secret, is that their friendship, like their loves, is false, thus proving that women are incapable of both friendship *and* love. At the end of the little narrative, Prévan and the three women's lovers pledge to each other infinite friendship ("On se jura amitié sans réserve") and the women, the false friends, are destroyed. In this subplot, virile friendship is sealed with the proof of feminine incapacity.

Friendship between the sexes may seem difficult and almost impossible, but the institution of best girlfriends also proves singularly ineffectual at warding off the ravages of intersubjective aggression, resentment, and exploitation. Madame de Volanges is fooled by Merteuil's friendship with her and ends up seeing her daughter destroyed as a consequence. It is in the name of friendship that Madame de Volanges offers Tourvel an early warning about Valmont. Before she does so, she establishes the bond of everlasting friendship that binds the two women to each other.[29] Friends may tell us the truth about our lovers, but their interventions, no matter how accurate, are hardly effective against the latter's charms:

> Even more false and dangerous than he is charming and seductive, he has never, since his early youth, done or said anything without an ulterior motive, and he has never had an ulterior motive that was not reprehensible. My friend, you know me. You know that of the virtues I aspire to, tolerance is the one I most value. If Valmont had been overwhelmed by the storms of passion, if like a thousand others, he were seduced by the errors of his age, I would condemn his behavior while taking pity on his person; I would wait in silence until the moment of his happy redemption when he would win again the respect of decent people. But Valmont is not like that. His behavior is the result of his principles. He knows exactly how far a man can go in indulging in horrible things without compromising himself and in order to be cruel and wicked without danger; he has chosen women as his victims.[30]

As the novel progresses, Valmont confirms the accuracy of Volanges's portrait of him, but the true friend's warning is powerless against a charming man.

Why doesn't Volanges's letter completely destroy all the reader's or Tourvel's interest in Valmont? The answer, we would have to say, lies somewhere in the limitations of enlightened friendship with regard to mystifying love. Tourvel's immunity to the truth indicates that she already loves. The most compelling of truths told in friendship have no power against love's lies of convenience. In this particular portrait of the lover, the truth is impotent. This novel offers a highly ambivalent portrait of Tourvel as a duped lover whose capacity for feeling is idealized as a function of her resistance to friendly warnings.[31] She begins to trust Valmont and feels that the strength of her virtue will be adequate as an instrument of his conversion.

Merteuil assumes a complicity between Valmont and herself, but his loyalty to her is guaranteed by the force of a threat. She supposedly is in possession of one of his secrets. Ideal friendship should exclude the use of force (in this case, blackmail), but Merteuil, unlike Tourvel, never seems to allow herself to enjoy such a comforting idea. If Valmont and Merteuil move between self-consciously ventriloquizing discourses of chivalry and gallantry and a rhetoric of "virile" friendship that is shared by warriors, it is because they both see their relations as struggles for domination. Later, when Merteuil accuses Valmont of only being able to act as either a slave or a tyrant in relationship to a woman, it comes as no surprise. This accusation comes deep into the novel, after Valmont has already seduced Tourvel and is demanding the fulfillment of his contractual agreement with Merteuil. Merteuil does not feel like giving in to these demands, and she "reads" him as being in love with Tourvel, insofar as he can be in love with any woman. She tells him that his love is neither pure nor tender, and it allows him to differentiate (temporarily) between women, by elevating one above all others only so that he can eventually enjoy her debasement. In loving a woman, he temporarily deludes himself into believing that she has a quality

> that puts her in a different class, and places all other women in a second order; this quality still attaches you to her [Tourvel], even when you offend her. This is how I imagine a sultan must feel for his favorite sultana: he is not prevented from sometimes preferring a simple concubine. My comparison seems all the more accurate to me because, like the sultan, you are never a woman's friend or a lover. You are always either her tyrant or her slave.[32]

His love of one woman makes him an enemy to all others. The master-slave relationship in which Valmont finds himself trapped is precisely the kind of relationship that Enlightened friendship wants to supplant and replace. The image of a tyrannical order is one that also contains the conditions for a fantasy of unfettered enjoyment.[33] The figure for the absolute impossibility of friendship between men and women is the seraglio, an Orientalized image that is fascinating and repugnant at the same time insofar as it memorializes for the European imagination both a more repressive order and a less fettered sensuality. In insisting on Merteuil's capitulation, Valmont only proves her to be correct, for he is despotic in his demands on women. He only understands force.

If Valmont is incapable of either love or friendship with a woman, then the only authentic relationship a woman can have with him is that of absolute enemy: this is Merteuil's political insight, and one for which she will suffer. Rousseau's Julie says to Saint-Preux, "Let's just be friends," and the novel deals with the consequences of the belatedness of this imperative. Julie writes to Saint-Preux: "I could believe that you were a fickle lover, but not a deceitful friend." Heterosexual friendship in *The New Héloïse* is supposed to ensure against passionate betrayals. Suddenly, for the first time, it seems possible to be friends with women. What is new about this sentimental and sublimated friendship that Julie proposes is that it offers one of modernity's first revisions of the notions of classical friendship.

If Derrida politicizes the problem of friendship, it is because he evokes, however obliquely, the question of feminine exclusion. In *The Politics of Friendship* he worries about whether or not women can be befriended, but in doing so he reframes the question of friendship altogether in a discussion of the declared and absolute enemy. Derrida concedes that women are traditionally excluded from the realm of canonical or classical friendship. He describes this exclusion as a double one: in the "great ethico-politic-philosophical discourses on friendship," there is "on the one hand, the exclusion of friendship between women and, on other hand, the exclusion of friendship between a man and a woman."[34] What kind of relationships can women have under such conditions? Women are restricted to the order of the master-slave relationship. When the enlightened friendship of the Republic is made available to all classes of society, it is presented as an order of revolutionary "fraternity." In order to rescue women from their double exclusion, one must be able, above all, to *befriend* them.[35] As we have seen, however, this is not so simple: what Laclos's novel anticipates is an open declaration of hostilities that we know now as the "war of the sexes." Friendship is only one more front to be negotiated. The question of feminine friends is set off and apart from the text in a long aside, which Derrida begins in this way:

> You will not, perhaps, have failed to register the fact that we are writing and describing friends as masculine—neuter-masculine. Do not consider this a distraction or a slip. It is, rather, a laborious way of letting a question furrow deeper. We are perhaps borne from the very first step by and towards the question: what is a friend in the feminine, and who, in the feminine is her friend?[36]

It is, of course, extremely important to note that a discussion of Nietzsche follows this allusion to the linguistic problem of feminine exclusion from the terms of masculine-neuter friendship.[37] It is no coincidence that Nietzsche's thinking should arise at such a moment, as he has been key in our reevaluation of feminism and femininity. Nietzsche understands the new respect for women as another one of Enlightened Democracy's mystifying and repressive moments: he points out that since the French Revolution, the influence of women has waned. For him, an open declaration of hostility is preferable to condescension and respect, to the gestures that allow women to be included in public life and in the workforce as an exalted "office boy" or assistant. Condescending inclusion is a new form of domination, particular to industrial/bureaucratic society, that has replaced military/aristocratic formations in which powerful women were able to exercise a kind of sovereignty.

In his 1988 article "The Politics of Friendship," which developed later into the book-length work with the same title, Derrida cites Nietzsche's exclusion of women from the realm of friendship in order to elaborate on his reading of Aristotle's "O my friends, there is no friend." For Derrida, the formulation of this exclusion is crucial to the politics of friendship. Friendship, it seems, is not available to anyone, but it is always only a virtuality, a promise of things to come. For Zarathustra, women do not know friendship: they only know love. In a footnote to the article Derrida comments on the Nietzschean formulation "Women are not yet ready for friendship":

> One must underscore here the "not yet," because it also extends to man (*Mann*), but first of all and once again to the "brother" of Zarathustra as to the future of a question, an appeal or a promise, a cry or a prayer. It does so in the performative mode of the apostrophe. There is as yet no friendship, no one has yet begun to think friendship. Nevertheless, in the experience of a sort of bereaved anticipation, we can already name the friendship that we have not yet encountered. We can already think that we do not yet have access to it. May we be able to do it. . . . Woman is not yet capable of friendship. But tell me, men, who among you is capable of friendship? . . . There is camaraderie: may there be friendship! . . . But since woman has not yet acceded to friendship because she remains—and that is love—either "slave" or "tyrant," the friendship to come continues to mean for Zarathustra: liberty, equality, fraternity. In short the model of a republic.[38]

Women have only love or hate at their disposal: men have friends and enemies. In classical friendship, it is women who are confined to the domain of the tyrant/slave, which is why there is no place for them in the Republic. The act of declaring oneself a friend of women is not enough to repair the effects of such an exclusion. To declare women suddenly capable of friendship and thus enfranchised citizens of the Republic is to neglect the residual resentment, a burdensome legacy that the excluded party must bear. When Merteuil asserts that she was born to exact revenge for her sex, she understands that what must take place is a continual struggle for power that no amount of inclusionary rhetoric can mitigate.

What must be excluded or repressed from representation is the subjugation of the weak in relationship to the strong in ancien régime France. It erupts in the center of the novel in the relationships of absolute domination between men and women. Enlightened relationships only repress the violent potential of every liaison in which one party is visibly weaker. Sadism is born with Enlightenment: it represents the danger of every liaison:

> At no other epoch has the weaker sex been treated with such respect by men as in ours; it is part of our democratic inclination and fundamental taste, exactly like our lack of respect for the aged. No wonder that such respect is immediately misused. One wants more, one learns to make demands, one feels the very degree of respect as practically an insult; one would prefer to compete for rights, in fact one prefers open warfare.[39]

Respectful inclusion is not enough, especially when we are involved with a struggle for the terms of heterosexual coexistence. Nietzsche reads feminism as a symptom of industrialization and modernization: one of its most crippling effects lies in the insufficiency of its demands. It demands friendship and respect for women under the conditions of total war between the sexes: the friendly man, then, the one who obeys his democratic inclination and accepts these terms, is much more dangerous than the contentious one, the one who is a competitor, a rival, an openly declared enemy. Nietzsche teaches us that an openly declared enemy is more valuable than a half-hearted friend.

This narrative anticipates the waning of the power of feminine sovereignty under the Enlightenment. Laclos's novel is situated at a transitional moment: the move from ancien régime to New Republic brings about the eclipse of Merteuil's power. The New Woman will be sentimentalized, domesticated, and respected. The democratic republics are slow to offer the vote to them, under the pretense of shielding the weaker sex from unsavory political concerns; everyone seems initially satisfied with the sequestration of women in domestic spaces. The will-to-power of women will have to be censured, and later an abstract principle of equality will try to nullify the power struggle even as it offers consolation prizes. When Merteuil declares war on Valmont, she brings the struggle of women into sharp focus and acknowledges the enmity that is the condition of her warlike coexistence with the opposite sex. The New Republic of friendship between the Enlightened libertines will not and does not repair the gap between the sexes: the final condition of sexual difference is all-out war.

Merteuil tries to rewrite the narrative of sexual difference; she cuts out sentimentality and must suffer the consequences of her reasonable, material understanding of the politics of a power struggle. Laclos is not so ambitious; he keeps the love story that he knows will sell. Merteuil does not so much mistake herself for the author as Laclos allegorizes, in her duplicity, the conditions of his own writing. In the artifice of his dissimulating posture can be seen the duplicitous writing techniques practiced by Merteuil herself. Producing a fiction is an ignoble affair: Laclos's duplicitous frame narrative is no more or less duplicitous than Merteuil's machinations. The Editor's Preface adroitly constructs an editor's task with an editor's—not an author's—modesty. The work *(ouvrage)* is described as a collection *(receuil)*: "Commissioned by the persons who came into possession of the letters and who I knew wanted them published, I was given the responsibility of putting the collection in order, and I asked for my pains nothing more than the liberty of deleting the ones that seemed useless."[40] This purported act of fictional suppression lends to the editor's task an austerity and modesty that are supported by the rest of the preface. Although our fictional editor was uncomfortable with the errors and unevenness of style in many of the letters, it is not his fault if they are found in the published correspondence. He was given strict limits as to his intervention. If he was unable to do entirely as he pleased with them, he takes the liberty of asserting that even if he were the one who wanted the letters published, he would hardly be able to hope for their success. The effacement of ambition, of will and intention as well as invention, leaves him with the humble tasks of the paper-pusher, just taking orders from some mysteriously superior force whose will is greater than his own. The frame of the framing fiction, however, throws into question the authority of the editor. In the Publisher's Note that precedes the preface, a complete disclaimer is offered: "We think it necessary to warn the public that, despite the title of this work and what the editor says in his preface, we cannot guarantee the authenticity of this collection; we have in fact good reason to believe that it is only a novel."[41] It would not be, then, the fault of the publisher if such a work were *just a fiction*. The ironic tone of the publisher's note is sealed by the statement that he finds it unbelievable that such bad people could actually populate this century of Philosophy and Enlightenment. The refusal to sign off on this work is a different version of Valmont's own signature excuse—"It was not my fault."

    The responsibility for the text is deferred by layers upon layers of disavowal. Fiction needs excuses. When Danceny offers Madame de Rosemonde the letters, he makes it clear that they will indubitably cast blame on one person, Merteuil.: "I believed that it would be rendering a service to society in unmasking a woman as dangerous as Madame de Merteuil, who is as you can see, the only and real cause of everything that occurred between M. Valmont and myself."[42] Valmont inscribes his signature, "It was not my fault," on all the letters of the collection as he puts the blame on Merteuil: the publication of the entire correspondence is constructed as his exoneration. Valmont may duel with Danceny, but his real opponent is Merteuil: by giving his rival the letters, Valmont gives him proof of the Marquise's guilt and, by proxy, his own innocence in the affair of Cécile. This is how Valmont "wins" the war. He is liberated of accountability and responsibility; the letters are proof that "it was not his fault" after all. Merteuil made him behave badly: she made him seduce Cécile and abandon Tourvel.

Danceny only wants to make public two letters: one to avenge himself and Valmont, and the other to exculpate Prévan (it was not his fault either). Merteuil is thus exposed to universal condemnation. She alone remains guilty at the novel's dénouement as all her machinations are exposed. The second letter published by Danceny proves that she is to blame in the matter of Prévan, who was entrapped by her intrigue and caught in her bedroom in the middle of what he believed to be a romantic tryst. Most critics have remarked on the reactionary nature of this ending in which the order of propriety is firmly reestablished. The overturning of such a suffocating system of ancien régime manners takes place not as literature but as revolution. Yet revolution produces machines and contraptions of its very own. In the case of the French Revolution, the guillotine serves as the radical editor of ancien régime worldliness; after this novel, the final cut is yet to come. The ceremonial game of form and protocol has everything to do with what de Man called "the arbitrary power play of the signifier" in a text that can "only be experienced as a dismemberment, a beheading or a castration."[43] Writers are always doomed to lose something, if not their heads.

Merteuil can be read as one of the most ambitious creations of an Enlightenment novelist. Her own ambition mirrors that of Laclos: she aspires to power by trying to rewrite herself out of the frame of representation, by erasing all traces of her machinations. The story of her failures ends up being both a cautionary tale and the model for the author's success. She is both copy and machine. Death is refused to her: her fate will be one of exile, marginalization, and disfigurement. Her condemnation does nothing, however, to prevent the proliferation on a mass scale of relationships of rationalization and instrumentalization. Just as the automaton's demise was prepared for by the waxing of the principle of automation, the eclipse of Merteuil's reason allowed the Enlightenment to keep a soft focus not only on relations between the sexes but also on the inexorable progress of reason into that realm. Her punishment is staged for the delectation of a violent sentimentality, which in order to disguise its banality as authenticity lays the blame for mechanical reproduction on the more reasonable party.

Notes

Introduction

1. Walter Benjamin, "Theses on the Philosophy of History," *Illuminations*, trans. Harry Zohn (New York: Schocken Books, 1969), 253–64.

2. Alfred Chapuis and Edmond Droz, *Automata: A Historical and Technological Study*, trans. Alec Reid (Neuchâtel: Editions du Griffon, 1958), 363.

3. The flash, here representing both flash of insight and momentary paralysis of thought, is the visual and temporal figure with which Benjamin represents historical materialism's ability to seize the significance of the past.

4. Benjamin, "Theses," 256.

5. Julien Offray de la Mettrie, "L'Homme-Machine," in *Oeuvres philosophiques* (Paris: Librairie Arthème Fayard, 1987), 54–118.

1. Doing It Like a Machine

1. Paul de Man, *Allegories of Reading: Figural Language in Rousseau, Nietzsche, Rilke, and Proust* (New Haven, Conn.: Yale University Press, 1979), 294.

2. "Nothing succeeds so well as detachment to produce fascination, provided that detachment be made to speak, as in the silence of the analyst." John Guillory, *Cultural Capital: The Problem of Literary Canon Formation* (Chicago: University of Chicago Press, 1993,) 186.

3. Paul Bénichou, *Man and Ethics: Studies in French Classicism*, trans. Elizabeth Hughes (Garden City, N.Y.: Anchor Books, 1971). *Morales du grand siècle* (Paris: Gallimard, 1948).

4. *Le Statut de la littérature: Mélanges offerts à Paul Bénichou*, ed. Marc Fumaroli (Geneva: Librairie Droz, 1982).

5. "Nous ne manions pas des corps, ni des mécaniques; nous écoutons, nous

Notes to Chapter 1

interprétons, nous confrontons des signaux et des volontés. Notre souci est de le faire avec le moins d'erreur possible; nos critères de vérité, nécessairement approximatifs et rarement pourvoyeurs de certitude, demandent à être maniés avec d'autant plus de soin et de rigueur, à l'abri de toute chimère qui prétende oublier l'homme." Bénichou, "Réflexions sur la critique littéraire," in *Le Statut de la littérature*, 20–21. All translations of Bénichou are mine.

6. We will examine the ways in which the term *theory* is characterized by a number of theory's most virulent critics in the work that follows.

7. Jean Molino, "Sur la méthode de Paul Bénichou," in *Le Statut de la littérature*, 22–35.

8. "Il fallait oublier le système comme tel et, tout en se maintenant dans la direction choisie, travailler sans préjugé." Bénichou, "Réflexions," 4.

9. In Peggy Kamuf's recent *The Division of Literature: The University in Deconstruction* (Chicago: University of Chicago Press, 1996), she discusses the debates in the French university about just this topic—the constitution of the object of literary studies in opposition or analogous to the object of science. This debate has wide and far-reaching consequences. Bénichou's claims for the literary object should be read and understood in the context of this conflict.

10. "The adjustment of critical practice to new socioinstitutional condition of literary pedagogy is registered symptomatically within theory by its tendency to model the intellectual work of the theorist on the new social form of intellectual work, the technobureaucratic labor of the new professional-managerial class. It is for that reason that the turn to rhetoric not only gestures beyond the narrow confines of the literary syllabus but also resurrects the ancient art of rhetoric as a technical practice, quite unlike either the 'art' of appreciation or the even more intuitive exercise of judgment or taste, the art of appreciation" (Guillory, *Cultural Capital*, 181). For a critique of Guillory's use of the term *theory*, see Kamuf's *Division of Literature*.

11. De Man, *Allegories of Reading*, 294.

12. "L'automate ne peut incliner l'esprit que parce qu'il y a communication constante entre la 'machine' corporelle et la pensée." Bénichou, *Morales*, 130.

13. Louis Van Delft, *Littérature et anthropologie: La nature humaine et caractère à l'âge classique* (Paris: Presses Universitaires de France, 1993).

14. "On ne doit parler, l'on ne doit écrire que pour l'instruction; et s'il arrive que l'on plaise, il ne faut pas néanmoins repentir, si cela sert à et à faire recevoir les verités qui doivent instruire." Jean de La Bruyère, *Les Caractères ou les moeurs de ce siècle*, ed. Robert Garapon (Paris: Classiques Garnier, 1990), 61–62. This and all other translations of La Bruyère are mine.

15. Jean Starobinski, *Blessings in Disguise, or the Morality of Evil*, trans. Arthur Goldhammer (Cambridge: Harvard University Press, 1993), 34.

16. Starobinski's analysis of the question of civilization is firmly based on a reading of Freud's theorization of the intense ambivalence surrounding this term.

17. "L'orateur et l'écrivain ne sauraient vaincre la joie qu'ils ont d'être applaudis;

mais ils devraient rougir d'eux-mêmes s'ils n'avaient cherché par leurs discours ou par leurs écrits que des éloges." La Bruyère, *Les Caractères*, 61.

18. "Dès lors, puisque le monde n'est plus que surface, 'la vraie substance' de l'écriture sera 'cette superficialité' par quoi le style se fait homogène à son objet. Face visible du sens, le style des *Caractères* vise ainsi à la représentation littéraire d'une réalité nouvelle." Jules Brody, *Du style à la pensée: Trois études sur* Les Caractères *de* La Bruyère (Lexington, Ky.: French Forum Publishers, 1980), 10.

19. Louis Van Delft, *Littérature et anthropologie*.

20. "Les roues, les ressorts, les mouvements sont cachés, rien ne paraît d'une montre que son aiguille, qui insensiblement s'avance et achève son tour: image du courtisan d'autant plus parfaite qu'après avoir fait assez de chemin, il revient souvent au même point d'où il est parti" (La Bruyère, *Les Caractères*, 243).

21. Walter Benjamin, *The Origin of German Tragic Drama*, trans. John Osborne (London: Verso, 1985), 97.

22. La Bruyère, *Les Caractères*, 242.

23. "L'unique signe et la seule marque de la pensée cachée dans le corps." All citations of Descartes are from the Pléiade edition of his complete works, *Oeuvres et Lettres*, ed. André Bridoux (Paris: Gallimard, 1953), 1320, and the translations are mine.

24. "Pour que la parole puisse servir à démarquer l'homme du règne animal, il faut en effet qu'elle s'exerce dans le cadre d'une communication authentique: qu'elle soit une opération de l'intelligence et de la volonté de celui qui parle, et qu'elle soit reconnue comme telle par celui qui écoute." Michael Moriarty, "La Parole dans *Les Caractères*," *Cahiers de l'Association Internationale des Etudes Francaises* 44 (May 1992): 279 (translation mine).

25. "N'espérez plus de candeur, de franchise, d'équité, de bons offices, de services, de bienveillance, de générosité, de fermeté dans un homme qui s'est depuis quelque temps livré à la cour, et qui secrètement veut sa fortune. Le reconnaissez-vous à son visage, à ses entretiens? *Il ne nomme plus chaque chose par son nom*; il n'y a plus pour lui de fripons, de fourbes, de sots et d'impertinents: celui dont il lui échapperait de dire ce qu'il en pense, est celui même qui l'empêcherait de cheminer; pensant mal de tout le monde, il n'en dit de personne. . . . il a une triste circonspection dans sa conduite et dans ses discours, une raillerie innocente, mais froide et contrainte, un ris forcé, des caresses contrefaites, une conversation interrompue et des distractions fréquentes. Il a des torrents de louanges pour ce qu'a fait ou ce qu'a dit un homme placé et qui est en faveur, et pour tout autre une sécheresse pulmonique" (La Bruyère, *Les Caractères*, 241).

26. Howard Caygill, "The Signficance of Allegory in the 'Ursprung des Deutschen Trauerspiels,'" in *1642 Literature and Power in the Seventeenth Century*, ed. Francis Barker et al. (Colchester: University of Essex, 1981), 211.

27. Benjamin, *Origin*, 79.

28. Moriarty , "La Parole," 279.

29. "Le sot est automate, il est machine, il est ressort; le poids l'emporte, le fait mouvoir, le fait tourner, et toujours, et dans le même sens, et avec la même égalité'; il est uniforme, il ne se dément point; qui l'a vu une fois l'a vu dans tous les instants et dans toutes les périodes de sa vie; c'est tout au plus le boeuf qui meugle, ou le merle qui siffle: il est fixé et déterminé par sa nature, et j'ose dire par son espèce. Ce qui paraît le moins en lui, c'est son âme: elle n'agit point, elle ne s'exerce point, elle se repose" (La Bruyère, *Les Caractères*, 343).

30. Descartes, "Lettre à Newcastle," 1256.

31. "MACHINE: se dit figurément en choses morales, des adresses, des artifices dont on use pour avancer le succès d'une affaire. Il a fait jouer toutes sortes de ressorts et de machines pour venir à bout de cette entreprise. Cet homme est grossier et pesant, c'est une machine, il ne sort point de sa chaise." Antoine Furetière, *Dictionnaire universel* (Geneva: Slatkine Reprints, 1970), t.2 (F–O). Originally published in 1690.

32. "En considérant la machine—animal ou automate—comme une imitation, ou plutôt, comme une mauvaise copie de l'homme, Descartes, sut sauvegarder das son système de la nature une hiéarchie métaphysique traditionnelle, qui garantissait l'autonomie humaine par rapport à la passivité de la chose mûe." Jules Brody, "Images de l'homme chez La Bruyère," *Esprit Créateur* 15, no. 1–2 (1975): 173.

33. "Les enfants des Dieux, pour ainsi dire, se tirent des règles de la nature, et en sont comme l'exception. Ils n'attendent presque rien du temps et des années. Le mérite chez eux devance l'âge. Ils naissent instruits, et ils sont plus tôt des hommes parfaits que le commun des hommes ne sort de l'enfance" (La Bruyère, *Les Caractères*, 107). La Bruyère's own annotation indicates here that the "children of the Gods" refer to the sons and grandsons of kings. The precocity of the royal children is a familiar theme of the time.

34. Starobinski, *Blessings*, 51.

35. Ibid., 45.

36. Moriarty, "La Parole," 280.

37. Franz Kafka, "An Imperial Message," in *The Penal Colony: Stories and Short Pieces,* rev. ed., trans. Willa and Edwin Muir (New York: Schocken, 1976), 158–60.

38. The ear as labyrinth is a figure that recurs in the work of Jacques Derrida. What is especially important for our purposes is his reading of Nietzsche, to be found in "Otobiographies: The Teaching of Nietzsche and the Politics of the Proper Name," in *The Ear of the Other: Otobiography, Transference, Translation: Texts and Discussions with Jacques Derrida,* trans. Avital Ronell and Peggy Kamuf; ed. Christie V. McDonald (New York: Schocken, 1985).

39. See Louis Marin on the architecture of Versailles and the topography of absolute power in "Classical, Baroque: Versailles, or the Architecture of the Prince," *Yale French Studies* 80 (1991): 167–82.

40. Starobinski, *Blessings*, 15.

41. "Des systèmes nouveaux de plus en plus ambitieux ont fait fureur, inspirés

eux aussi de disciplines et d'hypothèses étrangères à la littérature. Ces pérégrinations dépaysantes hors de la réalité des oeuvres s'étant données pour des méthodes, quiconque n'adoptait aucune desdites méthodes, ni la marxiste, ni la psychanalytique, ni la structuraliste sous quelqu'une de ses formes, tombait sous le soupçon de n'en avoir—lamentablement—aucune. En fait, on baptisait méthodes des vues systématiques préconçues touchant le réel, qui définissent déjà ou supposent défini ce qu'*est* la littérature: une projection déguisée de l'économie, l'expression inavouée de pulsions inconscientes, une organisation de formes ou de signes verbaux, etc. Les méthodes de travail qui accompagnent dans chacun de ces cas la théorie n'ont pas servi à l'établir; au contraire, elles résultent de cette théorie même; c'est un ensemble de procédés destinés à confirmer sa vérité, puisée à une source étrangère et imposées aux lettres comme un a priori. On ne devrait appeler méthode en critique littéraire, au sens strict de ce mot—voie d'approche vers une vérité, où ne soit pas présupposée la nature de cette vérité—que celle qui consiste à s'informer suffisamment, à manier correctement l'information et à interpréter de façon plausible, c'est-à-dire en évitant la région mentale où l'indémontrable et l'irréfutable ne font qu'un" (Bénichou, "Réflexions," 4–5).

42. David Lehman's *Signs of the Times: Deconstruction and the Fall of Paul de Man* (New York: Poseidon Press, 1991).

2. "What's the Difference?"

1. Sigmund Freud, "The Uncanny," in *Standard Edition of the Complete Psychological Works,* trans. James Strachey (London: Hogarth Press, 1953–1974), vol. 17: 216–52. The *Standard Edition* will be referred to henceforth as *SE.*

2. E. T. A. Hoffmann, "The Sand-Man," in *Selected Writings,* trans. Leonard J. Kent and Elizabeth C. Knight (Chicago: University of Chicago Press, 1969), 156.

3. David Ferris, "Aura, Resistance and the Event of History," in *Walter Benjamin: Theoretical Questions,* ed. David Ferris (Stanford, Calif.: Stanford University Press, 1996), 22.

4. Geoffrey Bennington, "Aberrations: De Man (and) the Machine," in *Reading de Man Reading,* ed. Lindsay Waters and Wlad Godzich (Minneapolis: University of Minnesota Press, 1989), 214.

5. Paul de Man, "Anthropomorphism and Trope in the Lyric," in *The Rhetoric of Romanticism* (New York: Columbia University Press, 1984).

6. Rodolphe Gasché, *The Wild Card of Reading: On Paul de Man* (Cambridge: Harvard University Press, 1998), 55.

7. Paul de Man, *Aesthetic Ideology* (Minneapolis: University of Minnesota Press, 1996), 82.

8. Gasché, *Wild Card,* 81.

9. "De Man was an eighteenth-century mechanical materialist, and much that strikes the postcontemporary reader as peculiar and idiosyncratic about his

work will be clarified by juxtaposition with the cultural politics of Enlightenment philosophes." Fredric Jameson, *Postmodernism; or, The Cultural Logic of Late Capitalism* (Durham, N.C.: Duke University Press, 1991), 246.

10. John Guillory, *Cultural Capital: The Problem of Literary Canon Formation* (Chicago: University of Chicago Press, 1993).

11. Lehman, *Signs of the Times*, 58.

12. "Psychoanalysis as Conspiracy Theory or One Aspect of the Feminist Fantasy," *International Journal for Clinical Psychoanalysis* 2, no. 2 (1996): 87–102.

13. "Assuming there is a Yale Mafia, then surely there must be a resident Godfather. One is forced to finger Paul de Man, who exhibits qualities that may earn him the role of Don Paolo, *capo de tutti capi*." Frank Lentricchia, *After the New Criticism* (Chicago: University of Chicago Press, 1980), 283–84.

14. David Lehman, "Deconstructing de Man's Life," *Newsweek*, February 15, 1988.

15. Lehman, *Signs of the Times*, 258–59.

16. De Man's 1955 letter to Poggioli concerned the rumors about his work for the Belgian newspaper as well as his management of Editions Hermès. See *Responses: On Paul de Man's Wartime Journalism*, ed. Werner Hamacher, Neil Hertz, and Thomas Keenan (Lincoln: University of Nebraska Press, 1989), 474–75.

17. Hendrik de Man's sympathy with the Germans was explicitly related to his hostility toward the capitalism of what he called the "plutodemocracies." The complications of the European political scene are dramatized in the confusion of left- and right-wing politics in Hendrik de Man's career.

18. Lehman, *Signs of the Times*, 200.

19. Richard Klein, "The Blindness of Hyperboles: The Ellipses of Insight," *diacritics* 3, no. 2 (summer 1973): 33–44.

20. Guillory, *Cultural Capital*, 201.

21. See Laurence Rickels, "Avuncular Structures," in *Aberrations of Mourning* (Detroit: Wayne State University Press, 1988).

22. There is yet another Hendrik de Man who is associated with de Manian guilt. He is Paul's son from his first, common-law marriage to Anaide Baraghian, whom Paul sent to Argentina and essentially abandoned there. Paul seems to be equally attached to leaving his son behind: losing Hendrik as brother, claiming him as father, and abandoning him as son, Paul never seems to be able to get close enough to or far enough away from the object of ambivalence. What is encrypted in Paul de Man's past is both the name of Hendrik and the name of psychoanalysis.

23. For a thorough discussion of the poverty of academic and intellectual dialogue in our times, see Jeffrey Wallen's *Closed Encounters: Literary Politics and Public Culture* (Minneapolis: University of Minnesota Press, 1998).

24. Rickels, *Aberrations of Mourning*, 142.

25. Sigmund Freud, "Introduction to Reik's *Ritual: Psychoanalytic Studies*," in *SE* 17:260–61.

Notes to Chapter 2

26. Jacques Derrida, "Signature, Event, Context," in *Margins of Philosophy*, trans. Alan Bass (Chicago: University of Chicago Press, 1982), 316.

27. Derrida, "Freud and the Scene of Writing," in *Writing and Difference*, trans. Alan Bass (Chicago: University of Chicago Press, 1978), 210.

28. Viktor Tausk, "Influencing Machine," in *Sexuality, War and Schizophrenia* (New Brunswick, N.J.: Transaction Publishers, 1991), 193.

29. See Laurence Rickels, "Psy Fi Explorations of Out Space: On Werther's Special Effects," in *Outing Goethe and His Age*, ed. Alice A. Kuzniar (Stanford, Calif.: Stanford University Press, 1996), 164.

30. See, for example, Jean Rousset on the Baroque in *L'Intérieur et l'extérieur: Essais sur la poésie et sur le théâtre au XVIIe siècle* (Paris: Corti, 1976).

31. See Jurgis Balsustraitis on the work of Salomon de Caus in *Anamorphic Art*, trans. W. J. Strachan (New York: Harry N. Abrams, 1977), 37–39.

32. Ibid., 66.

33. Lewis Mumford, *Technics and Civilization* (New York: Harcourt, Brace and Jovanovich, 1963), 111.

34. Dahlia Judovitz, "Vision in Descartes," in *Modernity and the Hegemony of Vision*, ed. David Michael Levin (Berkeley and Los Angeles: University of California Press, 1993), 70.

35. De Man deals at length with the rhetorical question in "Semiology and Rhetoric," the first chapter of *Allegories of Reading: Figural Language in Rousseau, Nietzsche, Rilke, and Proust* (New Haven, Conn.: Yale University Press, 1979).

36. Judovitz, "Vision in Descartes," 73–74.

37. In her essay *"Blade Runner's Moving Still," Camera Obscura* 27 (November 1991): 88–107, Elissa Marder makes an implicit argument for a reading of the film as an allegory of cinematic representation itself, especially in its use of the image of the "moving still" or photograph come to life.

38. "Il n'a pas encore perdu la douceur du miel qu'il contenait, il retient encore quelque chose de l'odeur des fleurs dont il a été recueilli; sa couleur, sa figure, sa grandeur, sont apparentes; il est dur, il est froid, on le touche, et si vous le frappez, il rendra quelque son" (*Oeuvres et lettres*, 279).

39. "Les paroles toutefois m'arrêtent, et je suis presque trompé par les termes du langage ordinaire; car nous disons que nous voyons la même cire, si on nous la présente, et non pas que jugeons que c'est la même, de ce qu'elle a même couleur et même figure: d'où je voudrais presque conclure, que l'on connaît la cire par la vision des yeux et non par la seule inspection de l'esprit, si par hasard je ne regardais par la fenêtre des hommes qui passent dans la rue, à la vue desquels je ne manque pas de dire que je vois des hommes, tout de même que je dis que vois de la cire; et cependant que vois-je de cette fenêtre, sinon des chapeaux et des manteaux, qui peuvent couvrir des *spectres* ou des *hommes feints* qui ne se remuent que par ressorts? Mais je juge que ce sont de vrais hommes et ainsi je comprends, par la

seule puissance de juger qui réside en mon esprit, ce que je croyais voir de mes yeux" (*Oeuvres et lettres*, 281; emphasis added).

40. See Jules Brody's analysis of "doubt" in "Images de l'homme chez La Bruyère."

41. There is another level of difference that is played out in the racial and ethnic mix of twentieth-century Los Angeles: it has to do with what Jeffrey Wallen has called a mongrelization of cultures and languages that produces an underclass that appear as background, as the "less than human," but not wholly identifiable swirl of extras. This dystopic vision of the metropolis has provided the setting for many film noir narratives where the question of racial discrimination and ethnic prejudice leaves its traces in the sense of anxiety and moral compromise that allude to off-scene violence and exploitation.

42. Judovitz, "Vision in Descartes," 83–84.

43. See Descartes's *Discours de la méthode* (1637), which concludes, "If there were machines that had the organs and the face of an ape or another animal deprived of reason, we would have no means of recognizing that they are not of the same nature as those animals; on the other hand, if there were machines that resembled our bodies, and imitated as much as possible our actions, we would always have two very sure methods for recognizing that these are not real men. The first is that they would never be able to use or compose either words or other signs, as we do in expressing our thoughts to others" (S'il y avait de telles machines qui eussent les organes et la figure d'un singe ou de quelque autre animal sans raison, nous n'aurions aucun moyen pour reconnaître qu'elles ne seraient pas en tout de même nature que ces animuax; au lieu que, s'il y en avait qui eussent la ressemblance de nos corps, et imitassent autant nos actions que moralement il serait possible, nous aurions toujours deux moyens très certains pour reconnaître qu'elles ne seraient point pour cela de vrais hommes. Dont le premier est que jamais elles ne pourraient user de paroles ni d'autre signes en les composant, comme nous faisons pour déclarer aux autres nos pensées; 164).

44. Kaja Silverman, "Back to the Future," *Camera Obscura* 27 (1991): 124–25.

45. Derrida's treatment of *Nachtrag* reminds us that the belatedness of memory formations has to do with the temporal oscillations inherent in writing and representation. See "Freud and the Scene of Writing."

46. Sigmund Freud, "Project for a Scientific Psychology," *SE* 1:281–397.

47. In *replication* and *reduplication* we find Deleuze's reading of the Leibnizian *pli*, which suggests the fold as a kind of machine proliferation, a machine in relationship to the infinite and a machine opposed to mechanical and Cartesian models. The Deleuzian machine is outside of the scope of this project but should be referred to as an alternative model of complexity at work. Gilles Deleuze, *The Fold: Leibniz and the Baroque,* trans. Tom Conley (Minneapolis: University of Minnesota Press, 1993).

48. Silverman, "Back to the Future," 124–25.

49. Jacques Derrida, "Cogito and the History of Madness," in *Writing and Difference.*

50. Michel Foucault, *Madness and Civilization,* trans. Richard Howard (New York: Random House, 1965).

51. "Et comment est-ce que je pourrais nier que ces mains et ce corps-ci soient à moi? Si ce n'est peut-être que je me compare à ces insensés, de qui le cerveau est tellement troublé et offusqué par les noires vapeures de la bile, qu'ils assurent constamment qu'ils sont des rois, lorqu'ils sont très pauvres; qu'ils sont vêtus d'or et de pourpre lorqu'ils sont tous nus" (Descartes, *Oeuvres et Lettres,* 268).

52. Perhaps it was so much uncertainty that was unacceptable to the Hollywood studio that produced the film. Ridley Scott was forced to release the film with a "happier ending." In the studio cut of the film, the last shots we see are of the couple safely enclosed in a hovercraft, skimming over an idyllic, natural landscape (with no signs of the technological sprawl of L.A. in sight, not even high-tension wires or suburban tract housing), accompanied by a voice-over of Deckard reassuring us that Rachel was a "special" model of Nexus 6 and that therefore she had no termination date. A Hollywood guarantee of immortality, then?

53. Derrida, "Freud and the Scene of Writing," 228.

54. Lehman, *Signs of the Times,* 268.

3. The Princess of Clèves Makes a Faux Pas

1. Nancy K. Miller, "Emphasis Added: Plots and Plausibilities in Women's Fiction," *PMLA* 96 (Winter 1981): 36–48.

2. Lehman, *Signs of the Times,* 261.

3. Odile Hullot-Kentor, "*Clèves* Goes to Business School: A Review of DeJean and Miller," *Stanford French Review* 13.2–3 (1989): 251–66.

4. Friedrich Nietzsche, *The Will to Power,* trans. Walter Kaufman and R. J. Hollingdale (New York: Random House, 1967).

5. One of the most obvious homologies between Nietzsche and La Rochefoucauld or La Bruyère is purely formal: the aphorism functions in all three cases as a critical and philosophical intervention. Aphoristic force has the power of naming.

6. Miller, "Emphasis Added," 47–48 (emphasis added).

7. Sigmund Freud, "Creative Writers and Day-Dreaming," in *The Standard Edition of the Complete Psychological Works,* trans. James Strachey (London: Hogarth Press, 1953–74), 9: 146–47 (emphasis added). This and all subsequent citations of Freud refer to the *Standard Edition,* hereafter cited as *SE.*

8. Freud, "Creative Writers and Day-Dreaming," 147.

9. Ibid.

10. In three footnotes to three different texts—"Three Essays on the Theory of Sexuality" (1905), "Character and Anal Erotism" (1908), and "Civilization and Its Discontents" (1930)—Freud points to the relationship between ambition and urination.

11. The power of the weak is a Nietzschean term if there ever was one: for Nietzsche, however, the power of the weak is not redemptive: it describes the triumph of Christianity. This debate cannot be thoroughly addressed here because of constraints of time and space. Let us say that it seems astonishing that such an idea can be suggested without mention of Nietzsche's critique of Christian metaphysics in which he shows that the virtue of the weak is a distorted moralizing and crippling mask that conceals nothing less or more than will-to-power. The triumph of Christian ethics is for Nietzsche a disaster for the Western World. We shall see that this "forgetting" of Nietzsche is extremely significant in the development of a feminist reading of *The Princess of Clèves*.

12. See, for example, Freud's "Three Essays on the Theory of Sexuality."

13. Sarah Kofman tries to show that Freud's theory of bisexuality is mostly a defensive position that he takes with regard to the question of femininity. She writes of "the purely speculative character of the masculine/feminine opposition. The thesis of bisexuality thus implies that Sigmund Freud himself could not have *been purely and simply* a man *(vir)*, that he could not have had *(purely)* masculine prejudices." Kofman, *The Enigma of Woman: Woman in Freud's Writings*, trans. Catherine Porter (Ithaca, N.Y.: Cornell University Press, 1985), 15.

14. Kofman, *Enigma*, 112–13. Freud's essay "Femininity" appears in *SE* 22:112–35.

15. Freud makes a point of insisting that he is talking about the writers of popular novels, not those acclaimed by critics.

16. Joan DeJean, "Lafayette's Ellipses: The Privileges of Anonymity," *PMLA* 99 (October 1984): 899.

17. For a more detailed description of the difference between "fore-pleasure" and pleasure, see part III, "Transformations of Puberty," in "Three Essays on the Theory of Sexuality" (*SE* 7:125–245).

18. It is possible to consider that convention or *les bienséances* are related to the Derridean notion of supplementarity.

19. "Bienséance n'est que le masque des vices; où la vertu règne, elle est inutile." Jean-Jacques Rousseau, *Complètes,* ed. Bernard Gagnebin (Paris: Bibliothèque de la Pléiade, Éditions Gallimard, 1964), vol. 2:424–25. All translations of Rousseau are mine.

20. Saint-Preux describes the much-commented-on kiss that Wolmar orchestrates between Julie and himself: "Alors prenant la main de sa femme et la mienne, il me dit en la serrant: notre amitié commence, en voici le cher lien, qu'elle soit indissoluble. Embrassez votre soeur et votre amie; traitez-la toujours comme telle; plus vous serez familier avec elle, mieux je penserai de vous. Mais vivez dans le tête-à-tête, comme si j'étois présent, ou devant moi comme si je n'y étois pas; voilà tout ce que je vous demande. Si vous préférez le dernier parti, vous le pouvez sans inquiétude; car comme je me réserve le droit de vous avertir de tout ce qui me déplaira, tant que je ne dirai rien, vous serez sûr de ne m'avoir point déplu."

"Il y avoit deux heures que ce discours m'auroit fort embarassé; mais M. de

Notes to Chapter 3

Wolmar commençoit à prendre une si grande autorité sur moi que j'y étois déjà presque accoutumé" (Rousseau, *Complètes*, 2:242–45).

21. Jean Starobinski, *Jean-Jacques Rousseau: Transparency and Obstruction*, trans. Arthur Goldhammer (Chicago: University of Chicago Press, 1988), 5.

22. "La bienséance n'interdit pas à Madame de Clèves d'aimer Nemours, elle lui interdit seulement de le montrer! . . . Madame de Clèves au contraire se révolte contre les bienséances qui sont dissimulation et mensonge; elle a vocation de transparence non de représentation, c'est à la sincérité qu'elle se sacrifie." Henri Coulet, *Le Roman jusqu'à la revolution* (Paris: A. Colin, 1968), 259. All translations of Coulet are mine.

23. Joan DeJean, *Literary Fortifications: Rousseau, Laclos, Sade* (Princeton, N.J.: Princeton University Press, 1984), 70.

24. "Le vide se creuse derrière les surfaces mensongères." Jean Starobinski, *L'Oeil vivant: Corneille, Racine, Rousseau, Stendhal* (Paris: Gallimard, 1975), 1. Translations are mine.

25. "Qu'il serait doux de vivre parmi nous, si la contenance exterieure etait toujours l'image des dispositions du coeur." Jean-Jacques Rousseau, *Discours sur les sciences et les arts*, in *Complètes* 3:7.

26. Walter Benjamin, *The Origin of German Tragic Drama*, trans. John Osborne (London: Verso, 1977).

27. Ibid., 74.

28. Domna Stanton, "The Ideal of 'Repos' in Seventeenth-Century French Literature," *L'Esprit Créateur* 15 (1975): 79–104.

29. "La magnificence et la galanterie n'ont jamais paru en France avec tant d'éclat que dans les dernières années du règne du Henri second" (241). This and all subsequent citations of Lafayette's fiction refer to the collection of her work *Romans et nouvelles*, ed. Émile Magne (Paris: Garnier Frères, 1961). Translations are mine.

30. "La blancheur de son teint et ses cheveux blonds lui donnaient un éclat que l'on n'a jamais vu qu'à elle" (248).

31. The most significant of these figures are the Princess of Montpensier and the Countess of Tende.

32. "La singularité d'un tel aveu, dont elle ne trouvait point d'exemple, lui en faisait voir tout le péril" (337).

33. DeJean, "Lafayette's Ellipses," 899.

34. Peggy Kamuf's *Signature Pieces: On the Institution of Authorship* (Ithaca, N.Y.: Cornell University Press, 1988) emphasizes the political and problematic aspects of institutions as they are inaugurated by signatures and acts of signing.

35. In *Terrible Sociability: The Text of Manners in Laclos, Goethe, and James* (Stanford, Calif.: Stanford University Press, 1993), Susan Winnett uses the image of masking to describe the mastery of *mondanité* in ancien régime France: "Worldly success involves creating for oneself an original version of a conventional mask" (19).

36. Sigmund Freud, *Civilization and Its Discontents*, in *SE* 21:59–145.

37. "L'ambition et la galanterie étaient l'âme de cette cour, et occupaient également les hommes et les femmes. Il y avait tant d'intérêts et tant de cabales différentes, et les dames y avaient tant de part que l'amour était toujours mêlé aux affaires et les affaires à l'amour" (252).

38. The tectonic shift that took place in the construction of the "modern" self can be mapped as a movement from externalized discipline to internalized self-discipline, from spectacular punishment to interiorized punishment. This latter shift is mapped out in Foucault's *Discipline and Punish: The Birth of the Prison*, trans. Alan Sheridan (New York: Random House, 1977). The concomitant internalization of codes of social behavior in classical and postclassical France has been described by Roger Chartier in *Lectures et lecteurs dans la France d'ancien régime* (Paris: Editions du Seuil, 1987). A power shift takes place in Early Modern France when the cultural superego is weakened as the internalized superego is strengthened: this interiority becomes the site of intrapsychic conflicts that are no less violent for their invisibility.

39. "Non, madame, . . . je ne me veux fier qu'à vous-même: c'est le chemin que mon coeur me conseille de prendre, et la raison me le conseille aussi. De l'humeur dont vous êtes, en vous laissant à votre liberté, je vous donne des bornes plus étroites que je ne pourrais vous en préscrire" (340).

40. "Mme de Chartres . . . faisait souvent à sa fille des peintures de l'amour; elle lui montrait ce qu'il a d'agréable pour la persuader plus aisément sur ce qu'elle lui en apprenait de dangereux; elle lui contait le peu de sincérité des hommes, leurs tromperies et leur infidélité, les malheurs domestiques où plongent les engagements" (248).

41. "Mais les hommes conservent-ils de la passion dans ces engagements éternels? Dois-je espérer un miracle en ma faveur et puis-je me mettre en état de voir certainement finir cette passion dont je ferais toute ma félicité?" (387).

42. "Vous êtes sur le bord du précipice: il faut de grands efforts et de grands violences pour vous retenir" (277–78).

43. Peggy Kamuf reads the effacement of the princess at the end of the novel as "the final effect of the repeated failure to differentiate between Mother and her constructions; the failure, in other words, to bury the parents remains" (96). Kamuf also reads the prince as one of the Mother's substitutes; after Madame de Chartres's death, he plays a pedagogical role, reinstills in the princess a sense of her duty, and finally, receives her confession.

44. "Comme il y avait beaucoup de monde, elle s'embarrassa dans sa robe et fit un faux pas: elle se servit de ce prétexte pour sortir d'un lieu où elle n'avait pas la force de demeurer et, feignant de ne se pouvoir soutenir, elle s'en alla chez elle" (348).

45. Jacques Derrida, *Parages* (Paris: Éditions Galilée, 1986), 28.

46. Sylvère Lotringer identifies Passion with Difference: "Difference never ceases to confront the rule with the irrepressible affirmation of the individual" (La Différence ne cesse, fondamentalement de faire entendre face à la Règle la reven-

dication irrépressible de l'individu). Sylvère Lotringer, "Le Roman impossible," *Poétique* 3 (1970): 300. The irreducibility of this Passion produces only a temporary disturbance in the world of what Lotringer calls the "Règle." The princess's retreat from society is a mark of Difference that does not indicate a break with or in social convention.

47. In *The Heroine's Text: Readings in the French and English Novel 1722-1782* (New York: Columbia University Press, 1980), Nancy K. Miller underscores that for the literary heroine "the danger of the dangerous relation is dependent on the logic of the faux-pas: in the politics of seduction, once proves generally to be enough. Thus the rule of female experience in male-authored fiction is the drama of a single misstep" (x). According to Miller's reading, the fatality and the singularity of the misstep are emphasized as a symptom of the male authorship and a certain narratological ideology.

48. "The being in a state of passion no longer knows himself and is no longer in control of his behavior. He acts on impulse, and despite himself, and when he goes over in his mind everything that he has said and done, it is only to confirm the fact that passion has grown like a gangrene, and that all spontaneous gestures were just so many involuntary confessions" (L'être en état de passion ne se connaît ni se conduit, il agit par impulsions et malgré soi, et quand il revit dans son esprit tou ce qu'il a fait, tout ce qu'il a dit ou n'aurait pas dû dire, c'est pour constater avec effroi que la passion a progressé comme une gangrène et que, chose pire encore, tous ces gestes spontanés furent autant d'aveux involontaires). Jean Rousset, *Forme et signification: Essais sur les structure littéraires de Corneille à Claudel* (Paris: Corti, 1962), 22. Translations are mine.

49. "Elle s'enferma seule dans son cabinet" (351).

50. "Poussé par le désir de lui parler, et rassuré par les espérances que lui donnait tout ce qu'il avait vu, il avança quelques pas, mais avec tant de trouble qu'une écharpe s'embarassa dans la fenêtre, de sorte qu'il fit du bruit" (367).

51. See "Sur *La Princesse de Clèves,*" in Michel Butor's *Répertoire I* (Paris: Minuit, 1960).

52. "Enfin, pour se donner quelque calme, elle pensa qu'il n'était point encore nécessaire qu'elle se fît la violence de prendre des résolutions; la bienséance lui donnait un temps considérable à se déterminer; mais elle résolut de demeurer ferme à n'avoir aucun commerce avec M. de Nemours" (391).

53. "M. le duc d'Orléans mourut, à Farmoutier, d'une espèce de maladie contagieuse. Il aimait une des plus belles femmes de la cour et en était aimé. Je ne vous la nommerai pas, parce qu'elle a même caché avec tant de soin la passion qu'elle avait pour ce prince qu'elle a mérité que l'on conserve sa réputation. Le hasard fit qu'elle reçut la nouvelle de la mort de son mari le même jour qu'elle apprit celle de M. d'Orléans; de sorte qu'elle eut ce prétexte pour cacher sa véritable affliction, sans avoir la peine de se contraindre" (59).

54. Walter Benjamin, "The Storyteller" in *Illuminations,* trans. Harry Zohn, (New York: Schocken Books), 87–88.

55. "Si vous jugez sur les apparences en ce lieu-ci, . . . vous serez souvent trompé: ce qui paraît n'est presque jamais la vérité" (263).

56. Howard Caygill, "Benjamin, Heidegger and Tradition," in *Walter Benjamin's Philosophy: Destruction and Experience,* ed. Andrew Benjamin and Peter Osborne (London: Routledge, 1994), 20.

57. Jacques Derrida, "Signature, Event, Context," in *Margins of Philosophy,* 328.

58. In his discussion of Benjamin's *Origins of German Tragic Drama,* Samuel Weber emphasizes the notion of "origin" as that which attempts "to reproduce and restore the unique." If this is extremely helpful in understanding Lafayette's novel, as the original modern novel in French, the relationship that he draws between the "extreme and the unique" helps us in understanding the princess herself as the first of many heroines of modernity. Samuel Weber, "Genealogy of Modernity: History, Myth and Allegory in Benjamin's *Origin of the German Mourning Play,*" *MLN* 106 (1991): 472.

59. DeJean's, Miller's, and Coulet's readings destroy the difference between seventeenth-century literary princesses and twentieth-century subjects seeking authentification of identificatory impulses. Rather than condemning these interpretations as "anachronistic," we can understand their destructive potential in a dialectical manner. Samuel Weber suggests that the "obliteration of the historical, which originates in the baroque and then continues in its modern reception, renders the historical significance of the baroque both constitutive of and at the same time inaccessible to modernity. The baroque mourning play thus begins to emerge as the origin of a modernity whose distinctive historicity resides, in part at least, precisely in the effacement of historical distinction" (480.)

60. Benjamin, "The Storyteller," 87.

4. Getting Ahead with Machines?

1. Walter Benjamin, "Theses on the Philosophy of History," in *Illuminations,* 253.

2. Andrew Benjamin and Peter Osborne write in their introduction "Destruction and Experience," which opens the volume of essays titled *Walter Benjamin's Philosophy: Destruction and Experience* (London: Routledge, 1994), "For Benjamin, 'destruction' always meant the destruction of some false or deceptive form of experience as the productive condition of the construction of a new relation to the object" (xi).

3. Benjamin, "Theses," 262–63.

4. Ursula Pia Jauch, *Jenseits der Maschine: Philosophie, Ironie und Ästhetik bei Julien Offray de La Mettrie (1709-1751)* (Munich: Carl Hanser Verlag, 1998). Jauch suggests that La Mettrie represents a secret, encoded Enlightenment discourse that challenged many of the basic intellectual assumptions of the philosophes.

5. "Instead of describing a mechanical man, he [La Mettrie] simply wants to

Notes to Chapter 4

demonstrate the possibility of explaining human beings by a single material. Moreover, it is not about claiming to be able to construct a model of this material being, or to dismantle it" (Au lieu donc de décrire un homme mécanique, il veut simplement démontrer la possibilité d'expliquer l'être human par la seule matière. Pour le reste, il ne s'agit nullement de prétendre construire un modèle de cet être matériel, d'en démonter les ressorts). Ann Thomson, "L'Homme-machine, mythe ou métaphore?" *Dix-huitième siècle* 20 (1988): 375. Translations are mine.

6. "S'il a fallu plus d'art à Vaucanson pour faire son *Fluteur*, que pour son *Canard*, il eût dû en employer encore davantage pour faire un *Parleur*; Machine qui ne peut plus être regardée comme impossible, surtout entre les mains d'un nouveau Prométhée; . . . le corps humain est une horloge, mais immense, et construite avec tant d'artifice et d'habileté, que si la roue qui sert à marquer ses secondes vient à s'arrêter; celle des minutes tourne et va toujours son train." Julien Offray de La Mettrie, "L'Homme-Machine," in *Oeuvres philosophiques* (Paris: Librairie Arthème Fayard, 1987), 109–10. Translations are mine.

7. "L'expérience et l'observation doivent donc seules nous guider ici. Elles se trouvent sans nombre dans les Fastes de Medecins, qui ont été Philosophes, et non dans le Philosophes, qui n'ont pas été Médecins. Ceux-ci ont parcouru, ont éclairé le Labyrinthe de l'Homme; ils nous ont seuls dévoilé ces ressorts cachés sous des enveloppes, qui dérobent à nos yeux tant de merveilles" (66).

8. "Si composée, qu'il est impossible de s'en faire d'abord une idée claire, et conséquemment de la définir" (66).

9. "C'est en vain qu'on se récrie sur l'empire de la Volonté. Pour un ordre qu'elle donne, elle subit cent fois le joug. Et quelle merveille que le corps obéisse dans l'état sain, puisqu'un torrent de sang et d'esprits vient l'y forcer; la volonté aiant pour Ministres une légion invisible de fluides plus vifs que l'Eclair, et toujours prêts à le servir! Mais comme c'est par les Nerfs que son pouvoir s'exerce, c'est aussi par eux qu'il est arrêté. La meilleure volonté d'un Amant épuisé, les plus violents désirs lui rendront-ils sa vigueur perdue? Hélas! non; et elle en sera la première punie, parce que, posées certaines circonstances, il n'est pas dans sa puissance de ne pas vouloir du plaisir" (103–4).

10. Françoise de Graffigny, *Lettres d'une péruvienne* (New York: Modern Language Association of America, 1993). All translations are mine. This edition includes introductions by Joan DeJean and Nancy K. Miller, who have done the crucial work of reinserting Graffigny's work in the history of the eighteenth-century novel by presenting this critical edition of her novel.

11. Jean-Jacques Courtine and Claudine Haroche, *Histoire du visage: Exprimer et taire ses émotions, XVIe–début XIX siècles* (Paris: Rivages, 1988.)

12. "Tels à peu près que certains jouets de leur enfance, imitation informe des êtres pensants, ils ont du poids aux yeux, de la légèreté au tact; la surface colorée, un intérieur informe; un prix apparent, aucune valeur réelle. Aussi ne sont-ils guère estimés par les autres nations, que comme les jolies bagatelles le sont dans la société.

Heureuse la nation qui n'a que la nature pour guide, la vérité pour principe, et la vertu pour premier mobile" (Graffigny, *Lettres*, 52).

13. Fénelon, *Traité de l'éducation des filles* (Paris: P. Auboin, 1687).

14. Carolyn Lougee, *Le Paradis des Femmes: Women, Salons, and Social Stratification in Seventeenth-Century France* (Princeton, N.J.: Princeton University Press, 1976), 79.

15. Elizabeth J. MacArthur, "Devious Narratives: Refusal of Closure in Two Eighteenth-Century Epistolary Novels," *Eighteenth-Century Studies* 21, no. 1 (1987): 1–20.

16. See René Girard, *Deceit, Desire and the Novel*, trans. Yvonne Freccero (Baltimore: Johns Hopkins University Press, 1961).

17. Irving Wohlfarth, "No-Man's-Land," in *Walter Benjamin's Philosophy: Destruction and Experience*, 157–58.

18. A. Doyon and L. Liaigre, *Jacques Vaucanson, mécanicien de génie* (Paris: Presses Universitaires de France, 1966). All translations are mine.

19. Vaucanson's automation-duck appears in Thomas Pynchon's picaresque and epic novel *Mason Dixon* (New York: Henry Holt, 1997). The novel is about Mason and Dixon, the eighteenth-century suveyors and their careers in the New World. The duck appears as an eloquent and lovestruck escapee from her maker's studios. Her pursuit of the "Erotick" leads her to haunt Armand, a French cook with whom she eventually falls in love. Armand leaves Paris to come to the New World in order to escape her and ends up joining Mason and Dixon's surveying team as a cook but the duck's longing and her love give her superpowers of "Flight" and "Invisibility" so that she is able to follow Armand to the New World. Finally, her metamorphosis is complete when the automation-duck makes the quick change from mechanical to Metaphysickal" and disappears: Mason speculates that she has become an angel, star, or celestial body.

20. "[La nourriture] y est digérée comme dans les vrais animaux, par dissolution, et non par trituration, comme le prétendent plusieurs physiciens; mais c'est ce que je me réserve à traiter et à faire voir dans l'occasion. La matière digérée dans l'estomac est conduite par des tuyaux, comme dans l'animal par ses boyaux, jusqu'à l'anus, où il y a un sphincter qui en permet la sortie." Citations of Vaucanson's letter are taken from the pamphlet "Le Mécanisme du flûteur automate" (Paris: Éditions des Archives Contemporaines, in conjunction with the Conservatoire des Arts et Métiers, 1985). All translations are mine. This pamphlet includes Vaucanson's paper on the flute player that was presented before the Académie Royale des Sciences as well as his 1738 letter to the Abbé Desfontaines. Both of these texts were published by Jacques Guérin and sold at the Hôtel de Longueville where the automatons were first displayed.

21. "Je ne prétends pas donner cette digestion pour une digestion parfaite, capable de faire du sang et des parties nourricières pour l'entretien de l'animal; on aurait mauvaise grace, je crois, à me faire ce reproche. Je ne prétends qu'imiter la

Notes to Chapter 4

mécanique de cette action en trois choses, qui sont (1) d'avaler le grain; (2) de le macérer, cuire ou dissoudre; (3) de le faire sortir dans un changement sensible" (19).

22. Strangely enough, rumor has it that Robert Houdin, master of tricks, escapes, and sleight of hand, discovered Vaucanson's trick late in the nineteenth century, before the automatons were lost in a fire.

23. "Tout se passe comme si le mécanisme en tant que système était mis en échec dans les sciences de la vie parce qu'il rend l'expérience impossible.... Mais cette mise en échec est aussi un succès: la théorie mécaniste, parce qu'elle se donne pour objet le modèle, le symbole et la métaphore, rend possible la constitution du vivant comme objet de science positive, expérimentale, limitée." Claire Salomon-Bayet, *L'Institution de la science et l'expérience du vivant* (Paris: Flammarion, 1978), 182–83. Translation mine.

24. "C'est lorsque Vaucanson truque qu'il est le plus savant—une caricature du savant d'autrefois. L'automate, devenue feinte expérimentale, nourrit ... un esprit d'enquête nouveau." Jean-Claude Beaune, *L'Automate et ses mobiles* (Paris: Flammarion, 1980), 237. All translations are mine.

25. "L'automate-porteur-de-son-principe-de-mouvement radicalise la machine géométrique—donne aux mouvements réels schématisées l'appendice d'une liberté quasi infinie. L'automate à cames de Vaucanson ou Jaquet-Droz permet enfin d'envisager les mouvements s'enchaînant et se reproduisant à l'infini, ou presque; on l'a vu, c'est à cet instant historique qu'il semble pourtant disparaître. La superposition du corps de connaissances et de pratiques automatiques au corps humain cesse d'être valide, n'est plus, en tout cas, explicitement revendiquée comme telle" (ibid., 257).

26. "L'automate est déplacé de ses fonctions analytiques qui continuent sans lui: l'automate disparaît en tant que modèle et objet solitaires pour qualifier la forme du travail en milieu industriel" (ibid., 256).

27. Doyon and Liaigre, *Jacques Vaucanson*, 143–45.

28. "Un dépôt public des modèles des machines principalement utilisées dans les arts et les fabriques." "Jacques Vaucanson" (Paris: Musée National des Techniques, 1983), 23.

29. "Un dépôt de cette espèce encouragera ceux qui se sentent du gout pour l'invention des machines; il excitera les capitalistes à former des speculations sur le produit des machines nouvelles" (ibid., 25).

30. In his *Confessions*, Rousseau relates how he, too, briefly harbored ambitions of making a fortune by touring and displaying a "fontaine d'Héron" (named after Heron of Alexandria, the great engineer of antiquity). His hopes were dashed after the fountain was broken and he was unable to repair it.

31. Doyon and Liaigre gathered their information about this deal from Colvée's personal papers, deposited with Madame de Savigny, Vaucanson's daughter, upon his death in 1750.

32. "En cas d'insuffisance de la somme, Vaucanson devait la parfaire. Le

remboursement de la somme avancée serait effectué, à Colvée, par prélèvement des deux tiers des premières recettes. Ensuite il toucherait, pendant six mois, quatre sols par livre (soit 20%) sur les résultats des représentations, à moins que ce laps de temps n'ait pas suffi au remboursement du principal de la somme avancée. Il était stipulé, en outre, que la machine finie ou non, resterait le gage des sommes due par Vaucanson à Colvée. Ce dernier se réservant le droit de retirer des mains des ouvriers les parties de la machine, ou la machine tout entière, en quelque lieu qu'elle se trouvât" (Doyon and Liaigre, *Jacques Vaucanson,* 21).

33. Ibid., 25.

34. Ibid., 26–27.

35. The reception from the Académie Royale des Sciences was initially very cold. In April 1738, there were 1,806 tickets sold during a period of 23–24 days. The price of a ticket was three pounds, or the weekly salary of a female laborer. Spectators were received in groups of ten to fifteen at a time, and Vaucanson himself would give an introduction to the performance, and the twelve airs in the repertoire of the Flute Player would be played. From the February 11 to May 30, 1738, the flute player grossed 17,000 pounds (ibid., 33–34.)

36. Ibid., 51.

37. See Voltaire's "De la nature de l'homme":

> Le hardi Vaucanson, rival de Prométhée,
> Semblait, de la nature imitant les ressorts,
> Prendre le feu des cieux pour animer les corps,
> Pour moi, loin des cités, sur les bords du Parnassez
> Je suivais la nature et cherchais la sagesse.

The flute player was hailed as a gift, then, a gift of fire, knowledge, enlightenment; the philosophes found their theories harmonized by the gracefully turned, popular airs that the automaton was able to play.

38. La Mettrie, *L'Homme-Machine,* ed. Gérard Delaloye (Holland: Jean-Jacques Pauvert, 1966).

39. "Jacques Vaucanson, s'étant appliqué dès sa jeunesse aux sciences, a consommé pour s'y perfectionner le peu de fortune qu'il tenait de ses pères. C'est dans cet état d'épuisement qu'il sentit l'impossibilité de mettre fin à des anatomies mouvantes qu'il avait commencées et qu'il songea à tirer du secours du produit de quelques machines capables d'exciter la curiosité du public, qu'il conçut le dessein de faire une statue jouant de flûte traversière avec embouchure et par l'action des doigts" (Doyon and Liaigre, 34).

40. "Il y travailla en effet et avec le peu qui lui restait et les emprunts qu'il a été obligé de faire, il est parvenu à finir cette machine dont le public connaît le succès, mais il en tirerait peu d'avantage si Sa Majesté ne le mettait à couvert d'un de ses créanciers; c'est le sieur Marguin qui, sous prétexte d'aimer les arts et les sciences et après avoir attiré le suppliant chez lui, a fait passer au suppliant deux actes aussi

illicites qu'onéreux. . . . Il est aisé de sentir combien ces clauses sont odieuses. Le sieur Marguin, moyennant 6 000 livres dont partie en logement et nourriture, veut absorber le produit d'une machine qui a coûté plus de 12 000 livres au suppliant qui, d'ailleurs, y a consacré ses talents et un travail assidu de plus de deux ans. Cependant, le sieur Marguin, pour soutenir son injustice, vient de le faire assigner au Châtelet de Paris, par exploit du 17 avril. Mais le suppliant espère que Sa Majesté, protectrice des Sciences et des Arts, ne permettra pas qu'il soit exposé à des poursuites, qui, en consumant son temps et les secours qu'il peut tirer de sa machine, l'empècheraient de suivre ses travaux et de se rendre utile au public et qu'il plaise à Sa Majesté, de lui pourvoir" (Doyon and Liaigre, 34–35; emphasis mine).

41. Martin Heidegger, *The Question of Technology and Other Essays*, trans. William Lovitt (New York: Garland, 1977), 167.

42. Fryer and Marshall cite and try to refute an 1832 article in which a D. Brewster writes on the automatons of the eighteenth-century engineers Henri Maillardet, Pierre and Henri-Louis Jaquet-Droz, and Vaucanson: "'Ingenious and beautiful as all these pieces of mechanism are, and surprising as their effects appear even to scientific spectators, the principal object of their inventions was to astonish and amuse the public.'" David M. Fryer and John C. Marshall, "The Motives of Jacques de Vaucanson," *Technology and Culture* 20 (April 1979): 257. Astonishing and amusing the public are not the goals of true scientists.

43. Doyon and Liaigre, *Jacques Vaucanson*, 41.

44. "Je fus hier matin à la Bibliothèque du roi avec Md. D.C., son Suisse et V. . . . J'ai vu beaucoup de manuscrits rongés de rats, qui auroient transporté un savant et qui ne m'ont rien fait du tout. Nous avons été au flûteur, qui m'a fait bien du plaisir." Françoise de Graffigny, *Correspondance de Mme de Graffigny*, ed. J. A. Dainard and Engligh Showalter (Oxford: Voltaire Foundation, 1989) 2:146. Translation is mine.

45. Vaucanson was introduced to Graffigny sometime thereafter. There is mention of Vaucanson in her correspondence as she tries, unsuccessfully, to convince him to go to Prussia when Frederick II offers him a position in his court in 1740.

46. "Il semble que le besoin de rendre la vie et son mouvement ait toujours préoccupé l'esprit de l'homme. . . . Peut-être même, dès l'instant où l'être humain façonna naïvement l'argile ou dégrossit le tronc d'arbre pour en faire une idole, eut-il l'idée de lui faire imiter quelques-uns de ses propres gestes." Alfred Chapuis and Edouard Gélis's study *Le Monde des automates* (Paris: Blondel La Rougery, 1928), 3. All translations are mine.

47. "Si chaque instrument pouvait sur un ordre donné ou même pressenti travailler de lui-même comme les statues de Dédale ou les trépieds de Vulcain qui se rendaient seuls aux réunions des dieux, si les navettes tissaient toutes seules, si l'archet [le plectre] jouait tout seul de la cithare; les entrepreneurs se passeraient d'ouvriers et les maîtres d'esclaves" (Chapuis and Gélis, *Le Monde*, 9).

48. "Les figures articulées primitives furent, pensons-nous, une des premières manifestations de l'art. L'homme, en imitant la nature, chercha à reproduire le

mouvement. Ce mouvement fut un plaisir à ses yeux (quand il ne lui inspira pas la crainte) et la représentation artificielle de la vie devint très tôt un divertissement populaire." Chapuis, *Automates: Machines automatiques et machinisme* (Geneva: S. A. des publications techniques, 1928), 12–13. Translations are mine.

49. Derek de Solla Price, in his article "Automata and the Origins of Mechanism and Mechanistic Philosophy," *Technology and Culture* 5 (Winter 1964), goes further than Chapuis: "We suggest that some strong innate urge toward mechanistic explanation led to the making of automata, and that from automata has evolved much of our technology, particularly the part embracing fine mechanism and scientific instrumentation" (10). This "innate urge" recalls Chapuis and Gélis's speculations about early man, naively fashioning images in clay. In one way or another, all histories of automatons try to "naturalize" the drive toward imitation.

50. Like Salomon-Bayet, Reed Benhamou describes the automaton as a potentially powerful and didactic model of the human body, in "From Curiosité to Utilité: The Automaton in Eighteenth Century France," *Studies in Eighteenth Century Culture* 17 (1987): "To develop a model that replicated vital human functions would increase medical competence; but it would even solve complex physico-technical problems. In an age largely convinced that the human body was an exceptionally intricate machine, the challenge of replicating that machinery was irresistible" (101).

51. "Le mérite de Vaucanson, c'est d'avoir, avec plus ou moins de succès mais obstination, opéré le déplacement d'intérêt et de méthode qui fait passer de l'automate-curiosité à la machine-outil" (Beaune, *L'Automate,* 259).

52. "En construisant des automates, Jacques Vaucanson ne voulait pas seulement créer des oeuvres qui étonnaient par leurs complications, qui suscitaient la curiosité des badauds. Son dessein était beaucoup plus ambitieux. S'aidant de son talent dans le domaine de la mécanique et de ses connaissances en anatomie, il désirait créer des êtres artificiels, des 'anatomies mouvantes.' Ces anatomies devaient reproduire, le plus fidèlement possible, les organes et les fonctions de l'être humain ou de l'animal. Leur véritable but n'était pas de divertir mais d'instruire et de favoriser les progrès de la médecine." Catherine Cardinal, introduction to *Le Mécanisme du flûteur automate* (Paris: Éditions des Archives Contemporaines, 1985), vii–viii. Translations are mine.

53. "Le désir qui anime [l'homme] à étendre ses connaissances, soit pour élever son esprit aux grandes vérités, soit pour se rendre utile à ses concitoyens." Denis Diderot and Jean d'Alembert, *Encyclopédie, ou dictionnaire raisonné des sciences, des arts et métiers*, 36 vols. (Bern and Lausanne: Société typographique, 1780), 4:578. Translations are mine.

54. "L'esprit accoutumé à la méditation, et avide d'en tirer quelque fruit, a dû trouver alors une espèce de ressource dans la découverte des propriétés des corps uniquement curieuse, découverte qui ne connaît point de bornes. En effet, si un grand nombre de connaissances agréables suffisait pour consoler de la privation d'une vérité utile, on pourrait dire que l'étude de la Nature, quand elle nous refuse le

nécessaire, fournit du moins, avec profusion à nos plaisirs: c'est une espèce de superflu qui supplée, quoique très imparfaitement, à ce qui nous manque." D'Alembert, "Discours préliminaire," 84. Translations are mine.

55. See Jacques Derrida's development of the notion of supplementarity in *Of Grammatology,* trans. Gayatri Chakravorty Spivak (Baltimore: Johns Hopkins University Press, 1976).

56. "Dans l'ordre de nos besoins et des objets de nos passions, le plaisir tient une des premières places, et la curiosité est un besoin pour qui sait penser, surtout lorsque ce désir inquiet est animé par une sorte de dépit de ne pouvoir entièrement se satisfaire. Un autre motif sert à nous soutenir dans un pareil travail; si l'utilité n'en est pas l'objet, elle peut en être au moins le prétexte. Il nous suffit d'avoir trouvé quelquefois un avantage réel dans certaines connaissances, où d'abord nous ne l'avions pas soupçonné, pour nous autoriser à regarder toutes les recherches de pure curiosité, comme pouvant nous être utiles. Voilà l'origine et la cause des progrès de cette vaste science, appelée en général Physique ou étude de la Nature, qui comprend tant de parties différentes" (d'Alembert, "Discours," 84).

57. See my discussion of pretexts in chapter 7, where it becomes obvious that friendship functions as a pretext for the pursuit of love and love as a pretext for the pursuit of erotic pleasures.

58. It seems important to remark here that Rousseau referred to his compulsive onanism as "ce dangereux supplément" that preserved his virginity while threatening his vigor. It seems that the pleasures of d'Alembert's curiosity also play an ambivalent, two-sided role: they compensate the thinker when he is frustrated, but they can completely lead him astray.

59. "Un mécanicien est celui qui tantôt applique aux machines un moteur nouveau, tantôt leur fait exécuter des opérations qu'on était obligé, avant lui, de confier à l'intelligence des hommes, on sait obtenir d'une machine des produits plus abondants et plus parfaits." Marie-Jean-Antoine Condorcet, "Èloge de Vaucanson," in *Oeuvres de Condorcet,* ed. A. Condorcet O'Conor and M. F. Arago, 12 vols. (Paris: F. Didot Frères, 1847–1849), 3:212–13.

60. Condorcet hails these innovations as signs of progress: Vaucanson was involved in a gradual process of perfecting silk production and, in so doing, saving human labor through rationalization and mechanization. He also praises the efficiency of Vaucanson's principles of automation and suggests that these developments eased the burden of workers by fragmenting their tasks (ibid., 3:219–20).

61. Doyon and Liaigre, *Jacques Vaucanson,* 202–3.

62. Condorcet, "Èloge," 11:206–7.

63. Heidegger, *The Question of Technology,* 13.

64. Samuel Weber, "Upsetting the Set Up: Remarks on Heidegger's Questing after Technics," *MLN* 104 (December 1989): 981.

65. Weber has discussed the problems of translation around Heidegger's use of the term *Technik* in English versions of *The Question of Technology.* Weber suggests

the use of the term *technics* instead of *technology*. Weber argues: "With regard to the German [Technik], the English word [technology] seems both too narrow and too theoretical. Too narrow, in excluding the meanings technique, craft, skill; and at the same time too theoretical, in suggesting that the knowledge involved is a form of applied science." Weber, "Upsetting the Set Up," 980–81.

    66. Heidegger, *The Question of Technology*, 22.

    67. Doyon and Liaigre, *Jacques Vaucanson*, 44.

    68. "Elle anima pendant dix ans le salon de son mari à Passy et à Paris, amis en 1748, sa liaison avec le duc de Richelieu amènera son mari de se séparer d'elle." Graffigny, *Correspondance*, 2:31.

    69. Ibid., 2:29.

    70. Doyon and Liaitre, *Jacques Vaucanson*, 223.

    71. Jean-François Marmontel, *Mémoires* (Clermont-Ferrand: G. de Bussac, 1972), 223. Translations are mine.

    72. "'Ah!! Monsieur,' s'écria-t-il, en se tournant vers La Poupelinière. 'Le bel ouvrage que je vois là! et l'excellent ouvrier que celui qui l'a fait! Cette plaque est mobile, elle s'ouvre, mais la charnière en est d'une délicatesse! Non il n'y a point de tabatière mieux travaillée. . . .' 'Quoi, Monsieur!' dit La Poupelinière en pâlissant. 'Vous êtes sûr que cette plaque s'ouvre?' 'Vraiment j'en suis sûr, je le vois,' dit Vaucanson, ravi d'admiration et d'aise. Rien n'est plus merveilleux.' 'Eh! que me fait votre merveille'" (ibid., 233).

    73. The story of the hinged door became notorious, and popular doggerel verse was composed in its honor: "You are warned that in Paris one can have made a fireplace on a spring, breaking through the house, where a lover can pass without anyone suspecting. One can see the machine at the home of a certain Farmer General. . . . Chez Madame de la P, who was the first to use it":

> Vous êtes avertis
> Qu'on fait fabriquer à Paris
> En perçant la maison
> Fonds de cheminée à ressort,
> Où l'amant peut passer le corps
> Sans que personne le devine.
> On pourra voir la machine
> Chez certain fermier général
> . . . .
> Chez Madame de la P,
> Qui s'en est servi la première.

(D'Estrée, cited in Doyon and Liaigre, *Jacques Vaucanson*, 224).

    74. Ibid., 223.

    75. Nancy K. Miller, *The Heroine's Text*, xi.

76. In marrying herself off, Des Hayes resembles other orphaned heroines of eighteenth-century fiction, above all, Richardson's Pamela and Marivaux's Marianne.

77. Marie Bonaparte, *The Life and Works of Edgar Allan Poe: A Psychoanalytic Interpretation*, rev. ed., trans. John Rodker (New York: Humanities Press, 1971); and Jacques Lacan, "Seminar on 'The Purloined Letter,'" trans. Jeffrey Mehlman, *Yale French Studies* 48 (1972): 39–72.

5. Don Juan Breaks All His Promises

1. For example, see John R. Searle's impatient criticisms of Derrida in "Reiterating Differences: A Reply to Derrida," *Glyph* 1, no. 1 (1977). Derrida's reply to Searle is published as "Limited Inc. abc," in *Limited Inc.*, trans. Samuel Weber (Baltimore: Johns Hopkins University Press, 1977). The debate between Searle and Derrida takes place around Derrida's reading and critique of J. L. Austin in "Signature, Event, Context."

2. Oscar Mandel, "The Legend of Don Juan," in *The Theatre of Don Juan: A Collection of Plays and Views 1630-1963* (Lincoln and London: University of Nebraska Press, 1963), 21.

3. Georges Poulet, *Studies in Human Time*, trans. Elliot Coleman (Baltimore: Johns Hopkins University Press, 1956), 100.

4. For David Ferris, the political significance of criticism resides in "the possibility of a criticism that is grounded in discontinuity," that cannot be sustained. Ferris goes on to explain the importance of discontinuity for the work of Walter Benjamin: "To isolate this moment of historicality or politics (the moment of the now of recognizability, the dialectical image, and the flash of lightning) is to repeat an essential phenomenology, in which the moment of shock (the shattering of the aura) will be preserved as the continuity of modernity." Ferris, "Aura and the Event of History," in *Walter Benjamin: Theoretical Questions*, ed. David Ferris (Stanford, Calif.: Stanford University Press, 1996), 25.

5. Carol Jacobs, "Allegories of Reading Paul de Man," in *Reading De Man Reading*, ed. Lindsay Waters and Wlad Godzich (Minneapolis: University of Minnesota Press, 1989), 108.

6. Walter Benjamin, "Theses on the Philosophy of History," 256.

7. Molière, *Oeuvres complètes*, 4 vols. (Paris: Garnier Flammarion, 1965). "O Ciel! que sens-je? un feu invisible me brûle, je n'en puis plus, et tout mon corps devient un brasier ardent. Ah!" (act 5, scene 6). All translations are mine and hereafter will be cited parenthetically by act and scene number.

8. Larry Riggs, "Ethics, Debts, and Identity in Don Juan," *Romance Quarterly* 34, no.2 (May 1987): 145.

9. In "La Notion de dette dans le Dom Juan de Moliere," *Revue d'Histoire du Theatre* 26 (1974), Michel Prunier notes that Don Juan is born as a theatrical

character when feudalism is about to disappear. The traces of feudalism's disappearance continue to haunt us and seem difficult to efface entirely.

10. "C'est une fort mauvaise politique que de se faire celer aux créanciers. Il est bon de les payer de quelque chose, et j'ai le secret de les renvoyer satisfaits sans leur donner un double" (4.3).

11. "Je crois que deux et deux sont quatre, Sganarelle, et que quatre et quatre sont huit" (3.3).

12. Sarah Kofman and Jean-Yves Masson, *Dom Juan ou le refus de la dette* (Paris: Galilée, 1991). All translations are mine.

13. Shoshana Felman, *The Literary Speech Act: Don Juan with J. L. Austin or Seduction in Two Languages*, trans. Catherine Porter (Ithaca, N.Y.: Cornell University Press, 1983).

14. J. L. Austin, *How to Do Things with Words* (Cambridge: Harvard University Press, 1975), 8–9.

15. *Dom Juan* (1.1).

16. In addition to the problem of theater, Vernet points out that to read Molière as the author of a corpus of texts is to commit an irremediable error (21). Molière's status as a classical author of a discrete corpus is highly problematic: he was also an actor and director of his own work. The force of his gestures and the particularities of his interpretations haunt the written traces of this theatrical work. See Max Vernet, *Molière: Côté jardin, côté cour* (Paris: A. G. Nizet, 1991).

17. Austin, *How to Do Things*, 21–22.

18. For Vernet, the question of parasitism is a crucial one in the theater of Molière: he takes up Michel Serres's work on Amphytrion and demonstrates that as a general principle, parasitism is not always a negative principle of doubling and mimesis.

19. Felman, *The Literary Speech Act*, 26–27.

20. "Je suis ami de Dom Juan . . . et je m'engage à vous faire faire raison par lui. Je m'oblige à le faire trouver au lieu que vous vous voudrez. . . . j'en réponds comme de moi-même" (3.3).

21. "Je suis votre serviteur, et de plus, votre débiteur. . . . Je vous prie encore une fois d'être persuadé que je suis tout à vous, et qu'il n'y a rien au monde que je ne fisse pour votre service" (4.3).

22. "Je prens à témoin l'homme que voilà de la parole que je vous donne. . . . Je vous réitère encore la promesse que je vous ai faite. . . . Voulez-vous que je vous fasse des serments épouvantables? Que le Ciel . . ." (2.2).

23. See Michel Serres on Don Juan in *Hermès I: Ou la communication* (Paris: Éditions de Minuit, 1968).

24. Austin, *How to Do Things*, 14.

25. Derrida, "Signature, Event, Context," 326.

26. Consider the importance of repetition in Artaud's writings on the theater,

and the Derridean analysis thereof. For Vernet, Artaud is the horizon of Molière's theater.

27. Derrida, "Limited Inc. abc," 89.

28. Christopher Braider, "The Semblance of the Dissembler" (unpublished manuscript).

29. Jacques Guicharnaud, *Molière: Une aventure théâtrale* (Paris: Gallimard, 1963), 72.

30. Lionel Gossman, *Men and Masks: A Study of Molière* (Baltimore: Johns Hopkins University Press, 1963), 37.

31. Nietzsche, *The Will to Power*, trans. Walter Kaufman and R. J. Hollingdale (New York: Random House, 1968), 135.

32. Riggs, "Ethics, Debts, and Identity in Don Juan," 141.

33. Norbert Elias, *The Court Society*, trans. Edmund Sephcott (New York: Pantheon, 1983).

34. For an excellent discussion of the notion of debt in Nietzsche's philosophy and Heidegger's reading of Nietzsche's translation of Anaximander, see Gary Shapiro's "Debts Due and Overdue: Beginnings of Philosophy in Nietzsche, Heidegger and Anaximander," in *Nietzsche, Genealogy, Morality*, ed. Richard Schacht (Berkeley and Los Angeles: University of California Press, 1994). Shapiro reminds us that "redemption" or *Erlösung* can refer to both debts and sins (365).

35. Max Vernet takes all this on by problematizing the very genealogy of Molière's work in relationship to the proper name of "Molière" as author. He is able to reactivate the significance of historical conditions (the status of the author in seventeenth-century France, the problem of defining the limits of Molière's corpus, and so on) through a theorization of the theatrical text as a figure of movement and *energeia*. Vernet manages to maintain a productive tension between his skepticism about historical contexts and theoretical explanation.

36. Friedrich Nietzsche, *The Genealogy of Morals*, trans. Francis Golffing (New York: Anchor Books, 1956), 189.

37. Monique Schneider also emphasizes the particularity of Don Juan's relationship to memory in *Don Juan et le procès de la séduction* (Paris: Aubier, 1994): "Don Juan est d'emblée celui qui refuse la mémoire. J. Rousset le qualifiera à juste titre d'amnésique" (19).

38. Nietzsche, *Genealogy*, 189–90.

39. "Comme *l'homme souverain,* il dispose du temps; s'il ne s'intéresse guère au passé (c'est Sganarelle qui est le comptable des conquêtes de Dom Juan, qui lui sert d'aide-mémoire), il anticipe les conquêtes futures, fait des projets pour enlever la paysanne qui lui résiste, même s'il est prêt, très vite, à abandonner une conquête pour une autre, et est voué au temps des rencontres successives, à l'émerveillement des rencontres, donc à un temps discontinu, où aucun moment n'est lié à un autre, ne répond de l'autre" (Kofman and Masson, *Dom Juan ou le refus de la dette,* 80–81).

Notes to Chapter 6

40. Otto Rank, *The Double: A Psychoanalytic Study,* trans. Harry Tucker Jr. (London: Maresfield Library, 1989).

41. Søren Kierkegaard, *Either/Or,* trans. George L. Stengren (New York: Harper and Row, 1986), 47.

42. Paul de Man, "Literary History and Literary Modernity," in *Blindness and Insight* (Minneapolis: University of Minnesota Press, 1971), 147.

43. "Eh! mourez le plus tôt que vous pourrez, c'est le mieux que vous puissiez faire. Il faut que chacun ait son tour, et j'enrage de voir les pères qui vivent autant que leur fils!" (4.5).

44. "Il est celui qui sait compter, et sur lequel on peut compter, l'homme mesurable dont la parole engage et qui est fiable. . . . c'est l'homme de la ratio, l'homme souverain, parfaitement maître de lui, ayant dompté. . . .l'homme de la pure dépense sans inhibition, ni restriction, ni *différance*" (Kofman and Masson, *Dom Juan ou le refus de la dette,* 80).

45. Paul de Man, "Literary History and Literary Modernity," 155–56.

46. Rodolphe Gasché, "Deconstruction as Criticism," *Glyph* 6 (1979): 209.

47. As a footnote to these reflections, I would like to append the following comments about the genealogy of literary theory and literary criticism. A certain critical forgetting takes place as well between applications of theories and readings of literature: total coverage is not possible. Derrida's work on the problematic place of literature in Austin's work seems to be occluded, if not completely ignored, in Shoshana Felman's compelling analysis of Don Juan. In a gesture that is surprisingly Don Juanesque, Felman makes light of Derrida's work by implying that he has no sense of humor, or at least that he has missed the sense of Austin's humor. Despite their shared references to Nietzsche, Sarah Kofman and Jean-Yves Masson in *Dom Juan ou le refus de la dette* mention Felman's *Literary Speech Acts* in a footnote, declare their sympathy for it, but otherwise banish this work to the margins as well. Max Vernet in *Molière: Côté jardin, côté cour* makes no references to either Felman's or Kofman and Masson's work on Don Juan; he is, on the other hand, conscientious about citing Derrida and Lacan in his discussions of textuality and identity. Criticism itself has achieved the monadic status of literary works. Even the work of those critics most resistant to theory can be included in this description: by ignoring theoretical insights and overturning notions of "progressive" Enlightenment in literary studies, they try to rewrite their own genealogies according to a logic of disjunction.

6. De Man on Rousseau

1. The phrases that are given as examples are left in French, with rough translations in parentheses: MACHINAL: Qui est fait sans intervention de la volonté, de l'intelligence, comme par une machine. V. Automatique, inconscient, instinctif, involontaire, irréfléchi, mécanique, réflexe. Un geste machinal. Réactions machinales.

MACHINATION: Ensemble de menées secrètes, plus ou moins déloyales. V.

Notes to Chapter 6

Agissement, complot, conspiration intrigue, manoeuvre, ruse. Ténébreuses, diaboliques machinations. Ourdir une machination.

MACHINER: Vielli. Former en secret (des desseins, des combinaisons malhonnêtes, illicites). V. Comploter, manigancer, ourdir, tramer; machination. Machiner un complot, une trahison. V. Conspirer, intriguer. Machiner la perte de qqn. From *Le Petit Robert* (Dictionnaire Alphabétique et Analogique de la Langue Française). The translations are mine.

2. Geoffrey Bennington, "Aberrations: De Man (and) the Machine," in *Reading de Man Reading*, ed. Lindsay Waters and Wlad Godzich (Minneapolis: University of Minnesota Press, 1989), 218.

3. See discussion in chapter 2. As has been already discussed, John Guillory makes an argument for the idea that de Manian practice not only inspired a powerful transferential fascination; it was also determined by the privileging of technical character of rhetorical reading, which combined with de Man's own enigmatic detachment as a professor was well-suited for the perpetuation of institutional conservatism. See Guillory, *Cultural Capital*, 234.

4. Neil Hertz, "Lurid Figures," in *Reading de Man Reading*, 102.

5. Jacques Derrida, "Freud and the Scene of Writing," in *Writing and Difference*, 196–231.

6. Victor Tausk, *Sexuality, War and Schizophrenia*, trans. Paul Roazen (New Brunswick, N.J.: Transaction Publishers, 1991),194.

7. Jacques Derrida, "Signature, Event, Context."

8. For Sarah Kofman, in "Le Respect des femmes" (1982), respect is a "machine," a "ruse qui permette d'introduire une distance, qui, en présence même de l'objet aimé et redouté, accorde du répit, sauve de l'imminence de la volupté et de la mort" (63).

9. Laurence Rickels and Avital Ronell have drawn on Tausk's work on schizophrenia and the machine in their analyses of the subject of technology. See Rickels's *The Case of California* (Baltimore: Johns Hopkins University Press, 1991), and Ronell's *The Telephone Book* (Lincoln: University of Nebraska Press, 1989).

10. Paul Roazen, *Brother Animal: The Story of Freud and Tausk* (New York: Knopf, 1969).

11. Diane Chauvelot, "Tausk: Sa mort comme transmission," *Ornicar?* 10 (June 1978): 3–26.

12. Tausk, *Sexuality, War and Schizophrenia*, 194.

13. See Joan Copjec's "The Anxiety of the Influencing Machine," *October* 23 (Winter 1982): 43–59, for further discussion of Tausk's study of Nataljia A. Copjec also notes the significance of Nataljia's dependency on writing as a form of communication.

14. Tausk, *Sexuality*, 195.

15. Freud, *SE* 12:73.

16. Rickels, *Case of California*, 87.

17. Starobinski, *Jean-Jacques Rousseau*, 10.

18. Ibid., 125.

19. In the opening pages of *Les Rêveries,* Rousseau claims that he has actually achieved a degree of unassailable tranquillity because the cumulative effect of all the attacks launched against him by his enemies is a sort of immunity to further distress. Physical suffering itself would only be a release for him.

20. Starobinski, *Jean-Jacques Rousseau,* 125.

21. "Quand je vis qu'à mon égard la raison était bannie de toutes les têtes et l'équité de tous les coeurs; quand je vis une génération frénétique se livrer tout entière à l'aveugle fureur de ses guides contre un infortuné qui jamais ne fit, ne voulut, ne rendit de mal à personne; quand après avoir vainement cherché un homme il fallut éteindre enfin ma lanterne et m'écrier: il n'y en a plus; alors je commençai à me voir seul sur la terre, et je compris que mes contemporains n'étaient par rapport à moi que des êtres mécaniques qui n'agissaient que par impulsion et dont je ne pouvais calculer l'action que par les lois de mouvement. Quelque intention, quelque passion que j'eusse pu supposer dans leurs âmes, elles n'auraient jamais expliqué leur conduite à mon égard d'une façon que je pusse entendre. C'est ainsi que leurs dispositions intérieures cessèrent d'être quelque chose pour moi; je ne vis plus en eux que des masses différemment mues, dépourvues à mon égard de toute moralité" (1:1077–107). All citations from Rousseau's work refer to the *Oeuvres complètes,* ed. Bernard Gagnebin and Marcel Raymond (Paris: Éditions Gallimard, 1959–1969). All translations are mine.

22. "Des carrières, des gouffres, des forges, des fourneaux, un appareil d'enclumes, de marteaux, de fumé et de feu succèdent aux douces images des travaux champêtres" (1:1067).

23. "Je ne saurais exprimer l'agitation confuse et contradictoire que je sentis dans mon coeur à cette découverte"(1:1071).

24. "Je me rappellerai toute ma vie une herborisation que je fis un jour du côté de la Robaila, montagne du justicier Clerc" (1:1070).

25. Barbara Johnson, "Rigorous Unreliability," *Yale French Studies* 69 (1985): 79.

26. "On voulut savoir où je l'avais pris. Je me trouble, je balbutie, et enfin je dis en rougissant que c'est Marion qui me l'a donné" (1:84).

27. "Elle était présente à ma pensée, je m'excusai sur le premier objet qui s'offrit" (1:86).

28. According to Cynthia Chase's reading of de Man on the subject of the "purloined ribbon," the effects of the "effet machinal" produce inexcusability in the form of what she calls "white lies." Chase, "Models of Narrative: Mechanical Doll, Exploding Machine," *Oxford Literary Review* 6 (1984): 45.

29. "Je répondis en rougissant jusqu'aux yeux que je n'avais pas eu ce bonheur. . . . On s'attendait à cette négative, on la provoquait même pour jouir du plaisir de m'avoir fait mentir" (1:1034).

30. "Il est donc certain que ne ni mon jugement ni ma volonté de dictèrent ma réponse et qu'elle fut l'effet machinal de mon embarras. Autrefois je n'avais point

Notes to Chapter 6

cet embarras et je faisais l'aveu de mes fautes avec plus de franchise que de honnêteté, parce que je ne doutais pas qu'on ne vit ce qui les rachetait et que je sentais au dedans de moi; mais l'oeil de la malignité me navre et me déconcerte; en devenant plus malheureux je suis devenu plus timide et jamais je n'ai menti que par timidité" (1:1035).

31. Starobinski emphasizes the ways in which Rousseau is never master of the spoken word. For Rousseau, the spoken word always leads to misunderstandings. Writing intervenes as a possibility for undoing the "malentendu." Rousseau exculpates himself for the lie that he utters: "Le langage ne va pas de soi, et Jean-Jacques n'est pas à son aise lorqu'il faut parler. Il n'est pas maître de sa parole, comme il n'est pas maître de sa passion. Il ne coïncide presque jamais avec ce qu'il dit: ses mots lui échappent, et il échappe à son discours" (Starobinski, *Jean-Jacques Rousseau*, 151).

32. This kind of denial could be called "dénégation," the French translation of the Freudian term *Verneinung*, or in English, *negation*. It is by means of a negation that one acquires knowledge of the unconscious; this often leads to the scandalous psychoanalytic interpretation of "no" as "yes."

33. "Je l'accusai d'avoir fait ce que je voulais faire et de m'avoir donné le ruban parce que mon intention était de le lui donner" (1:86).

34. De Man, *Allegories of Reading*, 294.

35. Freud, *SE* 12:63.

36. The structure of projection and displacement makes the Marion episode legible in a way that frees us from accusations of "misogyny" that Hertz alludes to in "Lurid Figures." Misogyny is too general a category here, and while allowing for a reading of ambivalence vis-à-vis femininity, it offers a rather limited idea of psycho-grammatical displacements.

37. There is not enough material to allow us to speculate on who might have been the object of Rousseau's homosexual desire; it is not insignificant, however, that book II of the *Confessions* includes a story of Rousseau's being molested at the hospice in Turin by an ardent "bandit" who called himself a Moor.

38. "J'ai procédé rondement dans celle [la confession] que je viens de faire, et l'on ne trouvera sûrement pas que j'ai ici pallié la noirceur de mon forfait. Mais je ne remplirais pas le but de ce livre si je n'exposais en même temps mes dispositions intérieures.... Jamais la méchanceté ne fut plus loin de moi que dans ce cruel moment, et lorsque je chargeai cette malheureuse fille, il est bizarre mais il est vrai que mon amitié pour elle en fut la cause"(1:86).

39. Carol Jacobs, "Allegories of Reading Paul de Man," 111.

40. Jacobs cites de Man from *Allegories of Reading*, 173.

41. "Cependant à ne consulter que la disposition où j'étais en le faisant, ce mensonge ne fut qu'un fruit de la mauvaise honte et bien loin qu'il partit d'une intention de nuire à celle qui en fut la victime, je puis jurer à la face du ciel qu'à l'instant même où cette honte invincible me l'arrachoit j'aurois donné tout mon sang avec joye pour en détourner l'effet sur moi seul" (1:1025).

42. De Man, *Allegories of Reading*, 281.

43. Ibid., 280.

44. "Madame de Vercillis ne m'a jamais dit un mot qui sentit l'affection, la pitié, la bienveillance.... Elle me jugea moins sur ce qu j'étois que ce qu'elle m'avait fait, et à forme de ne voir en moi qu'un laquais, elle m'empêcha de lui paroitre autre chose" (1:82).

45. Sandy Petrey, in *Speech Acts and Literary Theory* (New York: Routledge, 1990), 154, refers specifically to Steven Knapp and Walter Benn Michaels's "Against Theory," in *Against Theory: Literary Studies and the New Pragmatism*, ed. W. J. T. Mitchell (Chicago: University of Chicago Press, 1985).

46. Knapp and Michaels, "Against Theory," 23.

47. For a slightly different reading of de Man's attribution of emptiness to the name, and Rousseau's own attempts to direct a reading of his untimely ejaculation, one should consult Thomas Pepper's take on the excuse in relation to historicism and metonymy: "What is it to historicize something out of consciousness, to excuse oneself for not reading it, but to do so as Rousseau does in his guilty, ex post facto interpretations of his own uttering the name of 'Marion'? The metaphorical enchainment of the text is disrupted by this seizing on context.... A reading that reaches for extra-textual determinations in order to account for problems encountered in reading a text is one that avenges itself or expiates itself on the first object available. De Man is a theologian of the name." Pepper, *Singularities: Extremes of Theory in the Twentieth Century* (Cambridge: Cambridge University Press, 1997), 123.

48. Crébillon, *Les Égarements du coeur et de l'esprit* (Paris: Flammarion, 1985).

49. "Bonne fille, sage, et d'une fidélité à toute épreuve"(1:84).

50. De Man, *Allegories of Reading*, 296. For Freud in *Moses and Monotheism*, committing a murder is compared to mutilating a text (*SE* 23:7–137). Covering up the act is most important and most difficult in both cases.

51. "Un jour, j'étais à l'étendage dans la chambre de la calandre et j'en regardais les rouleaux de fonte: leur luisant flattoit ma vue, je fus tenté d'y poser mes doigts et je les promenais avec plaisir sur le licé du cylindre, quand le jeune Fazy s'étant mis dans la roue lui donna un demiquart de tour si adroitement qu'il n'y prit que le bout de mes deux plus longs doigts; mais c'en fut assez pour qu'ils fussent écrasés par le bout et que les deux ongles y restassent" (1:1036).

52. Chase writes, "The mechanical functioning of the work is the enjoyment concealed and practiced by the ruse of the aesthetic—by the mechanical coming into play of 'fiction,' of the 'excuse' that a signifying structure ... is the coming into play of the mechanical. The Rousseau essay evokes and indicts this enjoyment.... At the origin of the work, or art, would be the flaw of the mechanical, the disjuncture of performance from an intention or meaning. That primacy of the mechanical does violence both to meaning and to the body and language" (Chase, "Models of Narrative," 56–57).

53. "Quand la stérilité de ma conversation me forçait d'y suppléer par d'inno-

centes fictions, j'avais tort, parce qu'il ne faut point pour amuser autrui s'avilir soi-même; et quand entraîné par le plaisir d'écrire j'ajoutais à des choses réelles des ornements inventés j'avois plus de tort parce qu'orner la vérité par des fables c'est en effet la défigurer" (1:1038).

54. "Jamais la fausseté ne dicta mes mensonges, ils sont tous venus de faiblesse, mais cela m'excuse très mal. Avec une âme faible on peut tout au plus se garantir du vice, c'est être arrogant et téméraire d'oser professer de grandes vertus" (1:1039).

55. "The delusions of paranoiacs have an unpalatable external similarity and internal kinship to the systems of our philosophers" (Freud, SE 17:260–61).

56. For Freud, megalomania is another defensive mechanism within the paranoid system (like delusions of persecution and erotomania) that is designed to ward off the weight of the unbearable phrase "I love him." The enunciation of megalomania is "I do not love at all—I do not love anyone." And since, after all, one's libido must go somewhere, this proposition seems to be the psychological equivalent of the proposition "I love only myself." So this kind of contradiction gives us megalomania, which we may regard as a sexual overvaluation of the ego and may thus set aside the overvaluation of the love-object with which we are already familiar (SE 12:65).

57. "Voici le seul portrait d'homme, peint exactement d'après nature et dans toute sa vérité, qui existe et qui probablement existera jamais" (1:3).

58. "Pour du mal, il n'en est entré dans ma volonté de ma vie, et je doute qu'il y ait aucun homme au monde qui en ait réellement moins fait que moi" (1:1059).

59. Alexander Garcia Duttmann, "The Violence of Destruction," in *Walter Benjamin: Theoretical Questions,* ed. David Ferris (Stanford, Calif.: Stanford University Press, 1996), 184.

60. Cynthia Chase deals with precisely this difference in her essay titled "Trappings of an Education: Toward What We Do Not Yet Have," in *Responses: On Paul de Man's Wartime Journalism,* ed. Werner Hamacher, Neil Hertz, and Thomas Keenan (Lincoln: University of Nebraska Press, 1989). She also reads de Man on Rousseau in the context of de Man's work on Kleist's "Über das Marionettentheater."

61. Derrida describes the "violence and confusion" in de Man's discussion of "a solution to the Jewish problem" as "unforgivable." See "Like the Sound of the Sea Deep within a Shell," in *Responses,* 127–164. This discussion appeared in de Man's most notorious article, "Les Juifs dans la Littérature actuelle," published in *Le Soir* on March 4, 1941 (it is included in Paul de Man, *Wartime Journalism, 1939-1943,* ed. Werner Hermacher, Neil Hertz, and Thomas Keenan [Lincoln: University of Nebraska Press, 1989], 45). To have written of "solutions to a Jewish problem" is to have participated in the anti-Semitic rhetoric of National Socialism. In the other *Le Soir* articles, de Man uses the language of national identity and national spirit to compare German, French, Flemish, and Belgian cultures, not surprisingly in Germany's favor.

62. See Derrida, "Like the Sound of the Sea Deep within a Shell," 150.

Notes to Chapter 7

7. Friends

1. Georges May, *Le Dilemme du roman français au 18e siècle: Étude sur les rapports du roman et de la critique 1715-1761* (New Haven, Conn.: Yale University Press, 1963); Vivienne Mylne, *The Eighteenth-Century French Novel: Techniques of Illusion* (Manchester: Manchester University Press, 1965).

2. Andrew Benjamin, *Art, Mimesis and the Avant-Garde* (London: Routledge, 1991), 183–84.

3. "Le romancier a mauvaise conscience au XVIie siècle, le roman prétend toujours ne pas être un roman; il n'invente rien, il présente du réel à l'état brut." Jean Rousset, *Forme et signification*, 75.

4. "L'utilité de l'Ouvrage, qui peut-être sera encore plus contestée, me paraît pourtant plus facile à établir. Il me semble au moins que c'est rendre un service aux moeurs, que de dévoiler les moyens qu'emploient ceux qui en ont des mauvaises pour corrompre ceux qui en ont de bonnes, et je crois que ces Lettres pourront concourir efficacement à ce but." "Préface du rédacteur" to *Liaisons dangereuses*, in Choderlos Laclos, *Oeuvres complètes*, ed. Laurent Versini (Paris: Bibliothèque de la Pléiade, Editions Gallimard, 1979), 7. This and all subsequent translations of Laclos are mine; quoted passages hereafter will be cited by letter number.

5. "The discovery of the letter as a narrative medium, like the discovery of the movie camera in our era, had three principal appeals: documentary, narrative and formal." Janet Altman, *Epistolarity: Approaches to a Form* (Columbus: Ohio State University Press, 1982), 211.

6. Rousset describes the epistolary novel as a form that allowed for the fragmentation of perspective: the following description is saturated with figures of cinema: "This fragmentation of the point of view into different spaces allows for new and interesting effects: most of the epistolary novelists indulged in it; the successive and different projection of lighting on the same event or character is to be seen in Richardson, in Rousseau, and of course, in Laclos" (Cette fragmentation de l'optique en de multiples foyers permet des effets neufs et intéressants: la plupart des épistoliers s'y plaisent; la projection d'éclairages successifs et variables sur un même événement ou sur une même personnage s'observe chez Richardson, chez Rousseau et bien entendu chez Laclos). Rousset, *Forme et signification*, 86.

7. Theodor Adorno and Max Horkheimer, *Dialectic of Enlightenment*, trans. John Cumming (New York: Continuum Press, 1991).

8. Ibid., 83.

9. "The novel of letters still allows itself to be interrupted at various intervals, by long memoirs" (Le roman à lettres se laisse encore encombrer de longs mémoires intercalés). Laurent Versini, *Laclos et la tradition: Essai sur les sources et la technique des* Liaisons dangereuses (Paris: Klincksieck, 1968), 270.

10. Laurence Rickels, *The Case of California*, 188.

11. "The mental image evoked, then, by the Marquise de Merteuil may be

compared to that of a man's head on a woman's body, or, if one prefers, of a creature with a feminine figure and a masculine soul; and it must have been this perverse and frightening combination, even more than her méchanceté, machinations, or sensuality, that filled Laclos' contemporaries with the sensation of fascinated horror, of prurient loathing, which they were unable to define exactly, and which continues until now to make its impression." Aram Vartanian, "The Marquise de Merteuil: A Case of Mistaken Identity," *L'Esprit createur* 3, no. 4 (winter 1963): 176–77.

12. Ibid., 180.

13. "As we follow their complicity and witness the flawless evolution of their schemes, we become aware of the degree to which the system and code of an earlier novel of worldliness has been purified and refined into a perfect mechanism which they perfectly understand and govern. The erotic is in fact the domain in which the drive to dominance, power and freedom operates most flawlessly. . . . To regard someone as a purely erotic object is to reduce his psychology to the most mechanical and simplified elements, to make an already rigid code of psychological signs still more mechanistic. Indeed, to reduce social relations to erotic relations, human behavior to erotic comportment, as Valmont and the Marquise continually try to do, is to operate an important mechanization of social laws and human existence." Peter Brooks, *The Novel of Worldliness: Crebillon, Marivaux, Laclos, Stendhal* (Princeton, N.J.: Princeton University Press, 1969), 176–77.

14. "J'ai éprouvé plus d'une fois combien votre amitié pouvait être utile; je l'éprouve encore en ce moment; car je me sens plus calme depuis que je vous écris; au moins je parle à quelqu'un qui m'entend, et non aux automates près de qui je végète depuis ce matin. En vérité, plus je vis, et plus je suis tenté de croire qu'il n'y a que vous et moi dans le monde qui valions quelque chose" (Letter 100; emphasis added).

15. "Whether the Présidente attends to the meaning of the clichés or simply to their forms does not matter, for the letter's form, its very conventionality, is itself its meaning, and allows her, unlike Valmont, to see the truth about him and herself. . . . The constative, or cognitive function of the letter—what it says repeatedly about the hypocritical conventions . . . between the sexes—coincides with its performative function—what it does as it hides another truth behind a veil of clichés [the truth of Valmont's love]. Yet paradoxically, the letter's performative function illustrates its constative function only at the expense of contradicting it." Janie Vanpée, "Reading Differences: The Case of Letter 141 in *Les Liaisons Dangereuses*," *Eighteenth-Century Studies* 27, no. 1 (fall 1993): 106.

16. Austin, *How to Do Things with Words*, rev. ed. (Cambridge: Harvard University Press, 1975).

17. "A wager, a marriage and a declaration of war furnish a reasonable spectrum of the scope and impact of performative language." Sandy Petrey, *Speech Acts and Literary Theory*, 8.

18. "Elle dénote, surtout, une faiblesse de caractère presque toujours incurable

et qui s'oppose à tout; de sorte que, tandis que nous nous occuperions à former cette petite fille pour l'intrigue, nous n'en ferions qu'une femme facile. Or je connais rien de si plat que cette facilité de bêtise, qui si rend sans savoir ni comment ni pourquoi, uniquement parce qu'on l'attaque et qu'elle ne sait pas résister. Ces sortes de femmes ne sont absolument que des machines à plaisir. Vous me direz qu'il n'y a qu'à n'en faire que cela, et que c'est assez pour nos projets. . . . mais n'oublions que de ces machines-là, tout le monde parvient bientôt à en connaître les ressorts et les moteurs; ainsi que pour se servir de celle-ci sans danger, il faut se dépêcher, s'arrêter de bonne heure, et la briser ensuite" (Letter 106).

19. "Je sais fort bien que l'usage a introduit dans ce cas, un doute respectueux: mais vous savez aussi que ce n'est qu'une forme, un simple protocole; et j'étais, ce me semble, autorisé à croire que ces précautions minitieuses n'étaient plus nécessaires entre nous" (Letter 129).

20. "Je persiste, ma belle amie; non je ne suis point amoureux; et ce n'est pas ma faute si les circonstances me forcent d'en jouer le rôle" (Letter 137; emphasis added).

21. Not only its contents but its very delivery and reception are reported secondhand.

22. "By saying that the excuse is not only a fiction but also a machine one adds to the connotation of referential detachment, of gratuitous improvisation, that of the implacable repetition of a preordained pattern. . . . The machine is like the grammar of the text when it is isolated from its rhetoric, the merely formal element without which no text can be generated. There can be no use of language which is not, within a certain perspective, thus radically formal, i.e., mechanical." De Man, *Allegories of Reading*, 294.

23. Adorno and Horkheimer, *Dialectic*, 91.

24. Derrida, *Politics of Friendship*, trans. George Collins (London and New York: Verso, 1997).

25. She is trying to invoke what would later become an American postwar teenager phrase, "Let's just be friends," used by girls mostly to mediate relations between the sexes: the unwanted sexual advances of the other are managed by an ambivalent attempt at sublimation.

26. "Femme qui consent à parler d'amour, finit bientôt par en prendre" (Letter 127).

27. "Quittez donc ce langage que je ne puis ni ne veux entendre; renoncez à un sentiment qui m'offense et m'effraie. . . . Ce sentiment est-il donc le seul que vous puissiez connaître, et l'amour aura-t-il ce tort de plus à mes yeux, d'exclure l'amitié? . . . En vous offrant mon amitié, Monsieur, je vous donne tout ce qui est à moi, tout ce dont je puis disposer. Que pouvez-vous désirer davantage? Pour me livrer à ce sentiment si doux, si bien fait pour mon coeur, je n'attends que votre aveu; et la parole, que j'exige de vous que cette amitié suffira à votre bonheur. . . . Si comme vous le dites, vous êtes revenu de vos erreurs, n'aimeriez-vous pas mieux

être l'objet de l'amitié d'une femme honnête, que celui des remords d'une femme coupable?" (Letter 67).

28. "Cette amitié précieuse, dont sans doute vous m'avez cru digne, puisque vous avec bien voulu me l'offrir, qu'ai-je donc fait pour l'avoir perdue depuis? ... En effet, n'est-ce pas dans le sein de mon amie que j'ai déposé le secret de mon coeur?" (Letter 77).

29. "I have never doubted, my young and lovely friend, neither your friendship for me nor the sincere interest that you take in everything that concerns me. It is not in order to clear up this point that I hope to forever have a connection between us, that I respond to your response, but I feel that I must speak to you about Valmont" (Je n'ai jamais douté, ma jeune et belle amie, ni de l'amitié que vous avez pour moi, ni de l'intérêt sincère que vous prenez à tout ce qui me regarde. Ce n'est pas pour éclaircir ce point, que j'espère convenu à jamais entre nous, que je réponds à votre réponse: mais je ne crois pas pouvoir me dispenser de causer avec vous au sujet de Valmont; Letter 9).

30. "Encore plus faux et dangereux qu'il n'est aimable et séduisant, jamais, depuis sa plus grande jeunesse, il n'a fait un pas ou dit une parole sans avoir un projet, et jamais il n'eut un projet qui ne fût criminel. Mon amie, vous me connaissez; vous savez si, de vertus que je tâche d'acquérir, l'indulgence n'est pas celle que je chéris le plus. Aussi, si Valmont était entraîné par des passions fougueuses; si, comme mille autres, il était séduit par les erreurs de son âge, blâmant sa conduite je plaindrais sa personne, et j'attendrais, en silence, le temps où un retour heureux lui rendrait l'estime des gens honnêtes. Mais Valmont n'est pas cela: sa conduite est le résultat de ses principes. Il sait calculer tout ce qu'un homme peut se permettre d'horreurs sans se compromettre; et pour être cruel et méchant sans danger, il a choisi les femmes pour victimes" (Letter 9).

31. Tourvel, like Proust's protagonists/lovers Marcel and Swann, possesses in advance all the signs of the beloved's perfidy, the beloved's unworthiness, but like them she is unmoved. In *Proust and Signs* (New York: G. Braziller, 1972), Gilles Deleuze describes how Proust understood the truths of friendship as being on the side of the truths of philosophy. The truth of desire and of art is of an entirely different order; it is always arrived at in an untimely fashion, both too early and too late, and it is almost always painful. In the Proustian order, truth always has something to do with jealousy and suffering, and the absolute lack of reciprocity or symmetry. Friendship is that which would protect us from jealousy and suffering, but friendship is powerless in the face of love. Volanges's truth arrives both too early and too late.

32. "Qui la place dans une classe à part, et met toutes les autres en second ordre; qui vous tient encore attaché, même alors que vous l'outragez; tel que je conçois qu'un sultan peut le ressentir pour sa sultane favorite, ce qui ne l'empêche pas de lui préférer souvent une simple odalisque. Ma comparaison me paraît d'autant plus juste que, comme lui, jamais vous n'êtes ni l'amant ni l'ami d'une femme, mais toujours son tyran ou son esclave" (Lettre 141).

33. Earlier in the novel, Merteuil describes herself as an entire harem of different women during one of her nights of lovemaking with Belleroche.

34. Jacques Derrida, "The Politics of Friendship," *Journal of Philosophy* 85, no. 11 (November 1988): 642.

35. In the case of Rousseau and Marion, there is no reciprocity, no love or respect, merely a relationship of thwarted desire sublimated into a "bizarre" friendship that is no friendship at all. The desire, love, or attraction dominates the scene between Rousseau and his object, so much so that there arises nothing but confusion between Marion and himself. Rousseau was therefore not very successful at maintaining the terms of friendship with the young woman, despite the fact that earlier in his life he proved himself capable of sacrifices in the name of friendship with both Pleince and Fazy.

36. Jacques Derrida, *The Politics of Friendship*, 56.

37. For Derrida, friendship is always anticipatory, virtual—"not a given." It is a response to the Other, given in the name of responsibility. This responsibility is a delicate affair because friendship, as a Kantian Idea, must maintain an uneasy truce, "an unstable equilibrium" between love and respect. Love is the modality of attraction and fusion; respect keeps things away from each other and operates on the modality of distancing and separation.

38. Derrida, "The Politics of Friendship," 643.

39. Friedrich Nietzsche, *Beyond Good and Evil*, trans. Margaret Gowan (South Bend, Ind.: Gateway Editions, 1955), 166–67.

40. "Chargé de la mettre en ordre par les personnes à qui elle était parvenue, et que je savais dans l'intention de la publier, je n'ai demandé pour prix de mes soins, que la permission d'élaguer tout ce qui me paraîtrait inutile" (6).

41. "Nous croyons devoir prévenir le Public, que, malgré le titre de cet Ouvrage et ce qu'en dit le Rédacteur dans sa Préface, nous ne garantissons pas l'authenticité de ce Receuil, et que nous avons même de fortes raisons de penser que ce n'est qu'un Roman" (3).

42. "J'ai cru de plus que c'était rendre service à la société que de démasquer une femme aussi réellement dangereuse que l'est Mme. de Merteuil, et qui, comme vous pouvez le voir, est la seule, la véritable cause de tout ce qui s'est passé entre M. de Valmont et moi" (Letter 169).

43. De Man, *Allegories of Reading*, 296.

Index

Abbé Desfontaines: *Observations sur les écrits modernes,* 91
Académie Royale des Sciences (Academy of Science), xiii, 88, 93, 95, 105
Adorno, Theodor, 157, 160
Alembert, Jean d', 96; "Discours préliminaire," 96
Altman, Janet, 156
anachronism, 107, 109–11
anonymity: of Mme de Lafayette, 53, 58, 59, 65
Aristotle, 178; *Politica,* 94
Artaud, Antonin: *The Theater and Its Double,* 123
Aubignac, François d', 81
Austin, J. L., 112–15, 121, 144, 163–64

Balustraitis, Jurgis, 38
Baroque subjectivity, 53
Barthes, Roland, 13
Beaune, Jean-Claude, 86–87, 95
Bénichou, Paul, xii, 4, 6, 37, 51, 52; and literary criticism, 3–4, 19–20, 47; and the machine, 5–6, 15, 20, 128; *Man and Ethics: Studies in French Classicism,* 2; "Réflexions sur la critique littéraire," 3

Benjamin, Andrew, 155
Benjamin, Walter, 21, 73, 108, 152; and the court, 63, 65; and the courtier, 10, 11, 105; and historical materialism, ix-xi, 15, 24, 76–77, 80, 83, 85, 100, 126; and historicism, 34, 95, 109; and language, 13; and the novel, 71, 75; *The Origin of German Tragic Drama,* 10, 63; and subjectivity, 22; "Theses on the Philosophy of History," ix, 76; and *Trauerspiel,* 72
Bennington, Geoffrey, 128
Bertin, 100
bienséance, 53–54, 60–62, 66, 68–71, 147, 155, 159, 174
Blade Runner, 34–35, 38–40, 42–47
Blanchot: "Pas est la Chose," 69
Boileau, 124
Bonaparte, Marie, 104
Braider, Christopher, 117
Brody, Jules: and La Bruyère, 8, 12; and the machine in Descartes, 15
Brooks, Peter, 162
Buffon, 100
Bussy-Rabutin, 54
Butor, Michel, 70

Index

Cardinal, Catherine, 95
Castiglione, Baldassare, 9
Caygill, Howard, 73; and language in *Trauerspiel*, 72
Chapuis, Alfred, ix; *Le Monde des automates*, 94
Chase, Cynthia, 141
Chauvelot, Diane, 131
Colvée, Jean, 89, 90, 93
Condorcet, 99; "Éloge de Vaucanson," 98
Confessions (Rousseau), 138–42, 143–48, 150–51, 153, 169
Corneille: El Cid, 123, 124
Coulet, Henri: and *bienséance*, 61, 66; *Le Roman jusqu'à la révolution*, 61; reading of *The Princess of Clèves*, 62, 72, 73
Courtine, Jean-Jacques, 80
Coysevox, Antoine: *Berger jouant de la flute*, 88
Crébillon: *Les Égarements du coeur et de l'esprit*, 148
curiosity, 95–97

Dancourt, Mimi, 101
Dangerous Liasions (film), 157, 170–71
Dangerous Liaisons (novel), 148, 156–76, 177, 178–82
de Caus, Salomon, 37–38
deconstruction, xiii, 151, 152; criticism of by Lehman, 24–29, 32–33, 47, 48, 50; and literary criticism, 106–7
DeJean, Joan, 51, 52, 73; and the classical novel, 62; "Lafayette's Ellipses: The Privileges of Anonymity," 58, 64–65
de Man, Hendrik, 29
de Man, Paul, 1, 5, 22, 32, 40, 59; *Allegories of Reading*, 1, 137, 153; and Bénichou, 3, 20; criticism of by Lehman, 20, 24–25, 29; critique of anthropomorphism, 23, 24, 34; excuse as fiction, 169; hatred of, 48; idealization of by students, 30; journalism during WWII, 27, 32, 33, 151, 152; and language, 48, 141, 147, 148, 150; and literary criticism, 108, 152; and the machine, xi, 1, 2, 6, 20, 128; materialism, xi, 24; and modernity, 123, 124; and the problem of representation, 35; question of difference, xii, 39, 50, 128; reading of Rousseau, xiii, 129, 137–39, 141, 145–46, 153; as theoretical father, 30, 31; and writing as mutilation, 149
Derrida, Jacques, 31, 69, 130; "Biodegradables," 27; "Cogito and the History of Madness," 45; on de Man, 152, 153; and *différance*, 105; and Don Juan, 106; "Freud and the Scene of Writing," 36; and "heterosexual" friendship, 172; and language, 48; in Lehman, 25, 28, 32; "The Politics of Friendship," 178; *The Politics of Friendship*, 177; question of difference, xii; reading of Austin, 113, 115, 116; reading of Descartes, 45, 46; reading of Freud, 34; reading of Rousseau, 60, 96; and the signature, 73; as theoretical father of Klein, 30; and writing, 35, 46, 131
Descartes, René, xii, 20, 39, 40, 45; and the automaton, 44, 77, 136; *Discourse on the Method*, 12, 155; doubt, 42; and *esprit*, 41; and human body, 38; and the machine, 14, 15; *Meditations*, 37, 40, 45, 46
Des Hayes, Françoise-Catherine-Thérèse (Thérèse), xiii, 100–101, 102–5
Dick, Phillip K., *Do Androids Dream of Electric Sheep?* 34, 44

Index

différance, 35, 105
Dom Juan (Molière), xiii, 106, 109–12, 113–15, 116–24
Don Juan 106–26, 127
Doyon, A., 83–85, 88–90; *Jacques Vaucanson, mécanicien de génie*, 85
Droz, Edmund, ix
Duttmann, Alexander Garcia, 152

Ehrgeiz, 55, 56
Elias, Norbert, 67; *The Court Society*, 120
Encyclopédie, 88, 95, 96
Enlightenment, the, x, xi, xiii, 2, 157, 158, 171, 179, 182

Felman, Shoshana: and *Don Juan*, 112, 117, 121; and speech acts in *Don Juan*, 113–14, 118–19
feminist criticism, 51, 52; intellectual inhibitions in, 50; and *The Princess of Clèves*, 49, 53; and sexual difference, 59
Fénelon, François de Salignac de, 81; critique of worldliness, 82
Fleury, Minister de, 101
Foir Saint-Germain, 90
Fontenelle, 124
Forman, Milosz: *Valmont*, 157
Foucault: reading of Descartes, 45–46; *Madness and Civilization*, 45
Frears, Stephen: *Dangerous Liaisons*, 157, 170–71
Freud, 42, 48, 68, 154, 161; and ambition, 55, 56; and the automaton, 21, 22; *Civilization and Its Discontents*, 66; "Creative Writers and Day-Dreaming," 54, 58; "Der Dichter und das Phantasieren," 54; Essay on "The Uncanny," 21; and fantasies or daydreams, 54, 58, 59; "Femininity," 57; *Interpretation of Dreams*, 36;

and the machine, 104, 132; and paranoia, 34, 35, 129, 130, 134, 142, 151; "Project for a Scientific Psychology," 45; reading of, by Miller, 52, 56, 57; reading of Hoffman, 23; refusal to analyze Tausk, 131; "The Relation of the Poet to Day-Dreaming," 54; sexual difference in, 55; theory of bisexuality, 57
friendship, 171–79
Fryer, David M.: "The Motives of Jacques de Vaucanson," 92

Gasché, Rodolphe, 24; "Deconstruction as Criticism," 125
Gélis, Edouard: *Le Monde des automates*, 94
Godzich, Wlad, 128
Gossman, Lionel: *Men and Masks*, 118; reading of *Dom Juan*, 118–21
Graffigny, Françoise de, 103; account of Des Hayes in correspondence, 101; critique of worldliness, 82; *Lettres d'une péruvienne*, 80, 81, 83; reaction to Vaucanson's automaton, 93–94, 96
Grimm, 100
Guicharnaud, Jacques, 117
Guillory, John: on Paul de Man, 6, 24, 30–31

Haroche, Claudine, 80
Hecquet, *Traité de la digestion et des maladies de l'estomac suivant le système de la trituration et du broiement*, 84
Heidegger, Martin, 92; "The Question of Technology," 99
Hérault, Lieutenant, 92
Hertz, Neil: "Lurid Figures," 128
Hoffmann, E. T. A.: "The Sand-Man," 21, 23, 31

Index

Horkheimer, Max, 157, 160, 170
Hôtel de Mortagne, 88
Hullot-Kentor, Odile, 52, 59; "*Clèves* goes to Business School: A Review of DeJean and Miller," 51

impotence: as failure of will, 80
Invasion of the Body Snatchers, The, 26
invraisemblance: in *The Princess of Clèves,* 54

Jacobs, Carol, 109, 143
Jameson, Fredric, 24
Johnson, Barbara, 138
Judovitz, Dahlia, 40, 43, 48

Kafka, Franz: "Imperial Message," 17
Kakutani, Michiko, 33
Kant, 24
Kierkegaard, Søren: and *Don Juan,* xiii, 122, 123
Klein, Richard, 29–32; "The Blindness of Hyperboles: The Ellipses of Insight," 29
Kleist: *Über das marionettentheater,* 23, 139
Knapp, Steven, 146, 147
Kofman, Sarah, xiii; critique of Freud, 57; and Don Juan, 112, 116, 121; reading of *El Cid,* 123

La Bruyère: *Les Caractères,* 6, 8, 124; descriptions of the courtier, xii, 9–12, 14–15, 17–18, 26, 33, 137; and language, 44; and the machine, 6, 7, 15, 18; and *parole,* 12, 13; and pleasure, 8, 16
Lacan, Jacques, 104
Laclos, Choderlos de: *Dangerous Liaisons,* 148, 156–76, 177, 178–82
Lafayette, Madame de, 86; anonymity of, 53, 58, 59, 65; daydreaming as writing, 59; *The Princess of Clèves,* xii, 49–54, 58, 61–62, 63, 64–75

La Mettrie, Julien Offray de, x, 80; and the automaton, 78; *L'Homme-Machine,* 78, 79; and the machine, 79; and Vaucanson, 91
Lanson, Guy, 3; *Histoire de la littérature française,* 2
La Poupelinière, Monsieur Le Riche de, 90, 92, 100–103
La Rochefoucauld, 52, 60
Le Cat, 89
Lehman, David, 25–29, 32–34, 47, 48, 50–52; *Signs of the Times: Deconstruction and the Fall of Paul de Man,* xii, 20, 24
Le Soir (Belgium), 27, 29
Le Statut de la littérature: Mélanges offerts à Paul Bénichou, 3
Liaigre, L., 83, 84, 85, 88, 89, 90
Liebesstreben, 55, 56
Lougee, Carolyn, 82; *Le Paradis des Femmes,* 81
Louis XIV, 82
Louis XV, 88, 92

MacArthur, Elizabeth: "Devious Narratives: Refusal of Closure in Two Eighteenth-Century Epistolary Novels," 82
machiner, 129–130
Malkovich, John, 170
Mandel, Oscar: and Don Juan, 107–108, 111, 112
Marder, Elissa, 44, 45
Marguin, Jean, 90–93
marionette, 139
Marmontel, Jean-François, 102
Marshall, John C., "The Motives of Jacques de Vaucanson," 92
Marx, Karl, 98

Index

Masson, Jean-Yves: and Don Juan, 112, 116, 121; reading of *El Cid*, 123
May, Georges: ancien régime novel, 155
Mercure de France, 91
Michaels, Walter Benn, 146, 147
Miller, Nancy, 50–52, 73; "Emphasis Added: Plots and Plausibilities in Women's Fiction," 49–50, 54; and the feminine signature, 65; *The Heroine's Text*, 103; reading of Freud, 54–55, 56; and sexual difference, 57, 58
Molière: *Dom Juan*, xiii, 109–12, 113–15, 116–24; *Tartuffe*, 117
Molino, Jean: "Sur la méthode de Paul Bénichou," 3
Montessuy, 98
Moriarty, Michael: and language, 13, 18; and La Bruyère, 12, 17
Mozart: *Don Giovanni*, 122
Mumford, Lewis: *Technics and Civilization*, 38
Musée des Arts et Métiers, 99
Mylne, Vivienne: ancien régime novel, 155

Nachträglichkeit, 35, 45
New York Times, 33
Nietzsche, Friedrich, xii, 52, 119, 154, 177, 178; critique of Rousseau, 53; and the effects of language, 121; and feminism, 179; and Freud, 66; *The Genealogy of Morals*, 121; and indebtedness, 120; and masks, 60, 63; *On the Advantage and Disadvantage of History for Life*, 123; *The Will to Power*, 52

Orry, Jean, 88

paranoia, 34, 35, 129, 133, 142–43, 151
parole, 12, 13

Pascal, Blaise, 128; and the automaton, 6, 7
performative speech acts, 127, 144, 164
Perrault, Charles, 124, 125
Petrey, Sandy, 146, 147
Pfeiffer, Michelle, 170
Poe, Edgar Allen: *The Purloined Letter*, 104
poiesis, 99
Poulet, Georges: and Don Juan, 108, 112
précieuses, 82
Prince de Condé, 84
Princess of Clèves, The, xii, 49–54, 58, 61–62, 63, 64–75, 81, 118

Rameau, Jean Philippe, 90, 100, 101
Rank, Otto: study of Don Juan legend, 122
Richelieu, Duke of, 100, 101, 103
Rickels, Laurence, 31, 35, 37, 134, 159
Riggs, Larry: reading of *Don Juan*, 110, 111, 118, 119, 120, 121
Rimbaud, 123
Rivera, Geraldo, 25
Roazen, Paul, 131
Rousseau, Jean Jacques, 32, 101, 171, 172; and *bienséance*, 61; *Confessions*, 138–39, 143–44, 151, 153; *Discourse on the Sciences and the Arts*, 61; and dissimulation, 60–63; *effet machinal*, 161; and the excuse, xiii, 129, 144, 151, 169; *Julie: The New Héloïse*, 60, 82, 83, 156, 176; as outsider, 80; and paranoia, 135, 136, 142, 151; on possibility of ethical lying, 149; reading of by de Man, 129, 137–39, 141, 143, 145, 147, 148–50, 152–53; reading of by Derrida, 60, 96, 152–53; *Reveries of a Solitary Walker*, 130, 137, 142; ribbon incident, 138–48, 150, 169;

sentimentalization of virtue, 53; and writing, 134
Rousset, Jean, 155, 157; and Don Juan, 119; reading of *The Princess of Clèves*, 70

Salomon-Bayet, Claire: *L'Institution de la science et l'expérience du vivant*, 86
salons, 82
Saxe, Maréchale de, 100
Schneider, Monique: and Don Juan, 122
Scott, Ridley: *Blade Runner*, 34–35, 36–40, 42–47
Searle, John: critique of Derrida, 116
sexual difference, 128–29, 147–49, 160, 162, 179; and Freud, 54–56, 57–58; and literary criticism, 49–51, 59
Silverman, Kaja, 39, 44, 45
speech acts, 144
Stanton, Domna: and "repos," 64
Starobinski, Jean, 7; and *bienséance*, 61; and flattery in the court, 16, 17; and Rousseau, 134
Stendhal: *The Charterhouse of Parma*, 11

Tartuffe, 117
Tau Saxe, Maréchale de, 100
sk, Victor, 36, 130–34, 142
Technē, 99

Technik/technique, 87, 99
Tencin, Madame de, 101
Theophrastes: and flattery, 17
Thomson, Ann: and La Mettrie, 77
Tilghman, Christopher: "In a Father's Place," 28, 29
Trauerspiel, 13, 72

Valmont, xiii
Van Delft, Louis, 7, 8
Vartanian, Aram, 161, 162
Vaucanson, Jacques, xiii, 78–99, 102, 104, 105; *Mémoire descriptif*, 93
Vernet, Max, 113, 121
Voltaire, 90, 91, 100, 101
von Kempelen, Baron: automaton chess player, ix
vraisemblance, 53, 155

Weber, Samuel, 72; and *The Princess of Clèves*, 74; "Upsetting the Set Up: Remarks on Heidegger's Questing after Technics," 99
will-to-power, xii, 52, 66, 74; and *The Princess of Clèves*, 52, 73; renunciation as example of, 53; of women, 179
Wohlfarth, Irving, 83

Žizek, Slavoj, 37

Catherine Liu is assistant professor in the departments of Cultural Studies and Comparative Literature, French, and Italian at the University of Minnesota. She has published articles on psychoanalytic theory, Walter Benjamin, and feminist criticism and French literature of the ancien régime. She has also published widely in art criticism and curated exhibitions of contemporary art in New York and Los Angeles. She is the author of the novel *Oriental Girls Desire Romance,* and she contributed an essay to *Acting Out in Groups* (Minnesota, 1999), edited by Laurence A. Rickels.